Eco-Justice Press, LLC
P.O. Box 5409 Eugene, OR 97405
www.ecojusticepress.com

Eco-Justice: Essays on Theory and Practice in 2016

After-word by:
C.A. Bowers

Essays by:
C.A. Bowers, John A. Cassell, Nigora Erkaeva, Rolf Jucker, Ethan Lowenstein, Thomas Nelson, Joseph Progler, Chun-Ping Wang

Cover by David Diethelm, Eco-Justice Press
Photo by Quinsey Sablan

Library of Congress Control Number: 2016952934
ISBN 978-1-945432-02-6

Eco-Justice
Essays on Theory and Practice
in 2016

Eco-Justice Press

TABLE OF CONTENTS

A Note From The Publisher

This is the first *Theory and Practice* from Eco-Justice Press. We present to you a wide range of topics relating to eco-justice*, by authors from Europe, Asia, and the Americas. We are pleased to give the authors a new venue to present their thoughts and we appreciate their contributions. Also, we thank Chet Bowers for writing the After-word.

The idea for this book arose because it seemed a 'where are we now' perspective on how eco-justice principles are being thought about and practiced seemed useful—we hope you agree. This ongoing discussion will hopefully benefit all.

We intend the '2016' of this book to be followed by a '2017', and hope you will join the discussion, either as a reader or as an author. An email address for prospective authors to submit essays has been established. Essays will be reviewed for suitability for inclusion in the next publication. Authors should send a short introductory email, and their essay to:

Theory-and-Practice@ecojusticepress.com
(Check ecojusticepress.com for submission deadlines)

eco-justice - NOUN
The condition or principle of being just or equitable with respect to ecological sustainability and protection of the environment, as well as social and economic issues.
—Entry from British & World English dictionary

Developing a Language to Support Healthy Partnerships in Powerful Place-Based Education:
The Experience of the Southeast Michigan Stewardship Coalition

Ethan Lowenstein & Nigora Erkaeva

SEMIS is a visionary coalition that unites educators and community partners to grapple with the roots of social and ecological problems and generate practical, place-based applications that have the potential to transform education...Having worked extensively with coalitions of youth, faith communities, and non-profits to advance environmental sustainability and social justice, I had come to see coalition building as a challenging, delicate, and often arduous process, whose rewards are seldom greater than the effort invested. My experience with SEMIS has been an invigorating alternative.

—Justin Schott, Director of Ecoworks

The partnership models we introduce in this paper were born out of the necessity of forming partnerships in the messy, tumultuous, and often inspiring environment of K-12 schools and communities in the Detroit area during the early part of the 21st Century. As Justin Schott, a long-time member of the Southeast Michigan Stewardship Coalition (SEMIS Coalition) expresses above, there is something different about the SEMIS Coalition and its method of partnership formation. Indeed, the organization, now in its 8th year had the luxury of a strong foundation that brought together overlapping, while at the same time diverse perspectives. Its original founders and intimate members of the coalition since its inception included Rebecca Martusewicz (a leading teacher and scholar in EcoJustice Education), Susan Santone (Director of Creative Change Education Solutions), Shug Brandell (then

Director of the Michigan Coalition of Essential Schools), and Sister Gloria Rivera (Sister of the Immaculate Heart of Mary, long-time community leader in Detroit, and lead organizer of Detroit Bioneers). Accompanying these founding women were John Lupinacci (then Detroit teacher and EcoJustice scholar-activist) and Rebecca Nielsen (then a community partner with the National Wildlife Federation and who later served as the Coalition's Programming Director). As Lupinacci (2013) notes in a qualitative case study documenting the oral history of the founding and early years of the Coalition, each of the founders brought their own strengths to the vision of the organization and its view of partnerships as well as decades of organizing and teaching in the Detroit area. Accompanying the work of Dr. Martusewicz and her then graduate student, John Lupinacci, was the seminal influence of Chet Bowers. Around the time of the SEMIS Coalition's founding, Bowers and Martusewicz had been organizing academic ecojustice retreats that offered both professional development and scholarly opportunities for ecojustice educators to engage in "Revitalizing the Commons" in Detroit (Bowers & Martusewicz, 2006; Lupinacci, 2013).

This essay is authored by the current Director of the Southeast Michigan Stewardship Coalition (SEMIS Coalition) and a former graduate student member of its Steering Committee. While this story appears on paper as a single essay, it grew out of close to a decade of discussions about partnership that took part within the SEMIS Coalition's Steering Committee and between members of the Coalition (see, for example, Lowenstein, Frenzel & Schott, 2014).

The Southeast Michigan Stewardship Coalition

The SEMIS Coalition is a network of schools, community partners, and teacher educators that seeks to help young people to become citizen-stewards of their local communities and the Great Lakes Region (Ignaczak, 2014). The SEMIS Coalition is a hub of the Great Lakes Stewardship Initiative (GLSI). Three pillar practices that unite the GLSI are Place-based Education (PBE), sustained professional development, and community partnerships. Since 2008, SEMIS Coalition members have worked to pool our strengths to achieve what none of us could do alone. The Coalition is composed of teachers and students from 18 schools, more than 35 community partner organizations, and university teacher educators. Many of the schools in SEMIS are in the City of Detroit. Coalition staff and community partner members provide intensive and sustained opportunities for teacher personal and professional growth, including a 4-day summer institute and 5 days of whole coalition programming during the school year, as well as site-based curriculum coaching. In some cases, Coalition staff work with entire schools that are using a place-based approach to frame what they do. Such holistic, continuous, and locally contextualized opportunities for adult growth are necessary for educators to make changes to their practice (Gulamhussein, 2013). The Coalition is young, but already accomplished, especially in light of its modest financial resources and small planning team. In the first eight years, over 12,000 students have been involved in place-based stewardship activities.

Theories that Have Influenced the SEMIS Coalition's View of Partnership

The SEMIS Coalition was founded on two complementary approaches–Place-based Education (PBE) and EcoJustice Education (EJE). These approaches served to frame its ideas about partnership. Although definitions and descriptions of PBE vary (see, for example, Smith, 2002; Demarest, 2014; Smith & Sobel, 2010; Greenwood, 2013; www.glstewardship.org) at its most basic, PBE engages students in using their local communities to inquire into school subject matter. Place-based Education strengthens children's connection to others who live in their local region, as well as developing respect for their communities, members within those communities, and the environment around them (Sobel, 2005). Place-based scholars in the critical tradition (e.g., Grace Lee Boggs, Julia Putnam, David Greenwood, and Greg Smith) propose that PBE offers a vision of schooling in which students and communities are empowered to critique harmful social and economic structures and become "solutionaries" to create a more just world. When adults and youth engage in place-based learning together, learning is not something abstract, but closely related to their own identities, histories, and visions of a more just and healthy community and world. When educators approach PBE with a critical stance, the intersection between social and ecological issues of justice become obvious (Greenwood, 2016) and can become rich grounds for a formal analysis of the interconnected root cultural causes of these issues (Bowers, 2001; Martusewicz, Edmundson, & Lupinacci, 2015). An Ecojustice Education approach emphasizes this formal analysis of the cultural root causes of social and ecological crises and asks us to carefully examine the ways in which our cultural belief systems lead us to behave in ways that support or harm life. At its core, an EcoJustice Education views "community" as including both humans and other living things in the natural world and argues that the way to change our own cultural mindsets and enact practices that support life and caring is to change the language we use (the stories, metaphors, symbols, ways of speaking and listening) and the kinds of communities we form (Lowenstein, Martusewicz, & Voelker, 2010; Lupinacci, 2013; Martusewicz et al., 2015).

The language we use shapes how and why we partner. The stories we tell and the language we use shapes our behavior (Bowers & Flinders, 1990; Foucault, 1972; Martusewicz et al., 2015) and the purpose and qualities of our partnerships. Becoming critically conscious of the language we use to describe our relationships is a necessary condition to determining whether this language supports or harms healthy relationships (Bowers, 2006; Martusewicz et al., 2015; Plumwood, 1997). One of the strengths of PBE using an EcoJustice approach is its ability to help learners not only identify community practices and ways of thinking that have harmed them and their communities, but also those that have wisdom and guidance for how to create caring relationships in which knowledge and resources are shared freely without the exchange of money (Bowers 2001; Martusewicz et al., 2015). We have found that when young people and adults see the strengths of their communities through partnering with other community members and organizations, they start to identify beliefs and community practices based on caring and mutual aid. Once this happens, their individual self-concept and their vision and structures of their organizations and communities can change. This is especially important in urban and rural areas where young people and their adult guides are bombarded by media messages that devalue their communities and schools, and that support an ethic of competition not partnership.

Organizational Partnerships

In the SEMIS Coalition, we have tried to "walk the walk" and put into practice in our coalition community, the ways of being that we profess theoretically and pedagogically. We have come to see cultural change, not as a separate activity from forming partnerships, but as perhaps its most important result. As a coalition, we have developed over a hundred partnerships. Some have failed, faded away, or been superficial in nature, while others have become long-lasting and mutually transformational. Still others flared for a while and now sit dormant, waiting for the right moment in the future to reignite. Over the years we have learned to resist working with partners using a one-size-fits-all approach. In the last couple of years, we have analyzed partnership processes within their specific and local contexts and begun to look at scholarship from a variety of perspectives to see whether and how our ways of thinking about partnership are reflected there. We were curious if others were thinking along similar lines, and how we might benefit and add to the discussion.

Theory and Research on Phases of Organizational Partnership Development

The concept of organizational partnership has been widely discussed in a variety of fields and can be broadly defined as a relationship directed to achieve reciprocally set aims (Bringle, Clayton, & Price, 2009; Macdonald & Chrisp, 2005). One fertile site for research and thinking about partnership is in the field of Academic and Community Service-Learning. Enos & Morton (2003), for example, analyzing the complex reality of partnership development between campus and communities, argue for partnership that is democratic and process-based and that serves to create interdependence rather than mutual dependence. Based on their own experience, they theorize two phases of partnership development: transactional and transformational. The difference between these phases is that transactional partnership is directed to complete an agreed upon project without much emphasis on changing the structure and the aims of organizations. On the other hand, in the transformational partnership phase, organizations are open to uncertainty, aim to create larger meanings beyond the self-interest of parties, work toward sustained commitment, and are ready to change the structure of the organization to make positive change in the community beyond the project-based relations.

While Enos & Morton (2003) stress the non-linear and dynamic nature of partnerships between organizations and elaborate two partnership phases, Bringle, Clayton, & Price (2009), reviewing various articles on partnership, conceptualize three partnership phases: unilateral, transactional, and transformational. Similarly, Dorado and Giles (2004) address the development of service-learning partnership levels that are, tentative engagement, aligned engagement, and committed engagement. This theory of partnership aligns well with that proposed by Bringle et al. (2009). Both of these models recognize that at the initial stage of partnership development organizations see partnership as beneficial for themselves. The second stage of partnership is one in which there is a level of recognition that organizations are in an engaged relationship directed to meet both (not only a single) organization's needs. The third stage of partnership development sees this relationship not only meeting the needs of both organizations, but also stretching beyond project-based relationships. These analyses recognize that organizations could enter partnerships at different phases depending on their prior experiences. Most of the studies on partnerships to date have analyzed the nature of partnership development levels and the distinctive features that are crucial for successful partnership. As Dowling et al.'s (2004) literature review of the studies about the outcomes of successful partnerships showed, many studies have been done on various features of partnership, but studies that ground this theory in experience are limited. One of our aims is to add to, and create, a conversation with this literature through the theoretically grounded experiences we share here.

Perhaps some of the most exciting work on partnership that parallels EcoJustice Educational perspectives is emerging in the interlocking fields of economics (see, for example, Senge, Smith, Krushwitz, Laur, & Schley, 2008), leadership (Scharmer & Kaufer, 2007; Wheatley, 2007) and community studies and organizational theory (Block, 2009). We have found *An Other Kingdom: Departing the Consumer Culture* by Block, Brueggemann, and McKnight (2016), to be especially helpful in articulating a clear language of healthy and just partnerships. In a culture based on ideas of covenant as opposed to contract, "deliverables" of relationships are not specified and quantified, nor bound by exact measures of time. A culture of covenant supports "neighborly disciplines, rather than market disciplines" and is "held together by its depth of relatedness, its capacity to hold mystery, its willingness to stretch time, and endure silence" (Block et al., 2016, p. 6-7). Covenantal communities follow cultural norms of abundance, fallibility, and the common good. Silence, listening carefully, and reflection are valued. As in an EcoJustice approach to Place-based Education, these neighborly disciplines must be anchored in community traditions and routines, and necessarily must be based on an "affection for place" (Block et al., 2016, p. 77).

Below we share a model of partnership developmental phases among organizations synthesizing partnership theories in the literature. We further extend these partnership theories through articulating the distinctive features of partnership phases based on concrete experiences of partnership development by the SEMIS Coalition.

The Two Models

In order to authentically capture the experience of partnership development within the SEMIS Coalition, we created two thinking tools. One was born out of a need to discuss, explain, and reflect on where, why, and how we were dedicating organizational resources (such as time and money) (see Figure 1) in light of our eco-democratic mission. The second is a tool that helps us to understand the nature of particular partnerships, identify the possibilities and constraints within different partnerships, and set goals for increasing the degree and qualities of reciprocal and transformational partnerships (see Figure 2).

Partnership Operation Model

This model helps us reflect on the following questions: What is our orientation towards partnership?
Does this orientation align with our espoused goals?
Where do we focus our organizational energies and with what results?

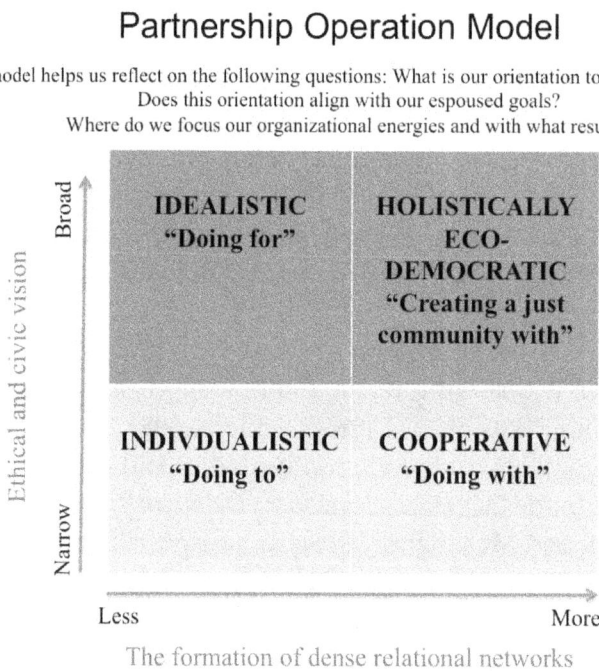

Figure 1   © 2016 SEMIS Coalition, Ethan Lowenstein, and Nigora Erkaeva

Our experience in building a coalition that draws its power from the presence of realistic and justice oriented partnerships in real life school and university contexts has taught us that it is important to devote focused energy to, and reflect on, at least two important orientations and their interactions: first, the development of a broad and collective ethical and civic vision, and second, the formation of dense relational networks. The above thinking tool not only helped us to address the way in which we allocate limited organizational resources, but also helped us better understand our own learning process, and explain the complexity with which we experienced partnership formation and navigation. The four-square model above locates partnership orientations as a product of resources and energy dedicated to activities along two continua - from a narrow to a broad vision and from sparse to dense relational networks. There are four fluid outcomes that are described inside these two arrows. They are individualistic, cooperative, idealistic and holistically eco-democratic.

Ethical and Civic Vision

The scope of an organization's ethical and civic vision depends on the degree of complexity with which it thinks about a set of core questions. The questions we highlight in the SEMIS Coalition when considering actions in community are based on our foundational approaches of Place-based Education and EcoJusice Education: Who is taken into account when making decisions? Who is included? Who is excluded? What beliefs and behaviors support life? What beliefs and behaviors harm life? Where do these beliefs and behaviors come from? How do we conserve and develop caring and democratic community practices that value diversity? How do we engage in the same kind of informed civic engagement in our own practice that we are promoting in our place-based educational approach? For example, an organization that defines and enacts its mission and values from a broad ethical and civic perspective, informed by an EcoJustice approach, emphasizes how issues, people, and beings in the more-than-human world are interconnected; views multiple perspectives as a strength in discussing and addressing complex problems; and views collaborative projects from a holistic perspective. We have begun to describe and share the specific materials we use in our collective learning and stories of our work on the SEMIS Coalition blog and website (www.semiscoalition.org) and readers may want to further investigate there.

The Formation of Dense Relational Networks

When organizations partner, they create the potential for individuals within those organizations to form deep and trusting relationships with each other. When large numbers of these potential relationships actually form, social network theorists call the social networks "dense" (http://www.the-vital-edge.com/what-is-network-density/). Transforming potential relationships to actual relational networks takes time and energy and organizations must purposefully and strategically allocate resources there. Wisdom, ways of thinking, values, skills, and resources can flow more easily across dense relational networks, and cultural capital can thus accrue within and between organizations when such networks are present (Putnam, 2001). In the last 15 years, there have been some exciting developments in methods for the rapid formation of ethical communities with large numbers of people by such authors as Peter Block (2009), Peter Senge, Otto Scharmer and colleagues (Senge, Scharmer, Jaworski, & Flowers, 2005), and Henri Lipmanowicz and Keith McCandless (2016). Such theory and research indicates that when people have a common and broad ethical and civic vision, the development of dense clusters of relationships can be accelerated with group processes and discussion protocols deliberately designed to be consistent with this vision.

Partnership Orientations

One of our realizations in the SEMIS Coalition has been that it's difficult to make sense of the formation of ethical and civic vision apart from the formation of dense relational networks. Individualistic actions result from a narrow vision in the context of sparse efforts to build relationships between organizations (see Figure 1). In this scenario, organizations act upon or "do to" others in a partnership. When partnering organizations have narrow visions that are strictly confined to their own organizational missions while also failing to dedicate resources and time to the development of trusting and caring relationships, each organization ends up seeking to meet its own needs without considering the needs of the other organizations. In our model this is called an individualistic orientation.

In contrast to an individualistic orientation, partnering organizations may have a cooperative orientation and get to know each other quite well, dedicating tons of resources to this relationship building. If their ethical and civic visions are shallow, however, they will fail to transform structures and commonly held beliefs to more democratic forms. Organizations in the cooperative orientation that are "doing with," will develop their relationships within bounded projects that are not placed within a broader social and ecological context.

A third orientation is idealistic. Here partnering organizations may have a very broad vision that holds promise in the abstract, but cannot or do not put this vision into practice because commitments of time, energy and resources have not been dedicated to the formation of the kinds of relational networks that enact and collectively build on this vision in a sustained way. The result of this "doing for" orientation is lots of initial energy around an important vision that ends up going nowhere and ultimately breeds cynicism about both the vision and the partnership.

In partnerships where we have a broad vision that is built on dense efforts from both parties, we move toward a fourth orientation termed holistically eco-democratic that considers the vision of organizations in a larger social and ecological context. In this orientation, the broadest ethical and civic vision allows organizations to consider not only human-to-human relationships, but also human to non-human relationships (Lupinacci, 2013; Lupinacci & Happel-Parkins, 2016; Lupinacci & Happel-Parkins, 2015). Few organizations are able to fully operate in this partnership space because concepts of "civics," "democracy," and "citizenship" in education have been largely anthropocentric (Lowenstein, 2012; Lowenstein, 2014). When partnering organizations both operate with holistically eco-democratic orientations, "partnership" becomes an inadequate and constraining descriptor, and "membership" in community replaces the idea of partnership. In this community space, the web of interconnected relationships with humans, the natural environment, and other species is included in all collaborative actions. Here, members are devoted to achieving this vision with intense effort, at the same time that incredible energy and resources are released for common and flexible use. Because "organizations" are made up of individuals who come and go, and are located within a powerful culture, and often rapidly changing institutional structures, even when a community is formed between organizations in which members feel a deep sense of belonging and share a common vision, vigilance is required to collectively adapt and change as the context shifts. Eco-democratic communities that operate within our current culture of hierarchy do not "achieve" eco-democracy, rather they enter a state of it and must continually act to form and maintain this state. Thankfully, the effort to remain in an eco-democratic state can become less over time as community rituals and routines are identified, created, reinforced, and maintained.

Our second mental model is an attempt to articulate essential components of different orientations in the partnership process that organizations experience over time and the patterns of qualities that partnerships may reflect along the journey. Both of our models complement each other, while at the same time provide different visual tools to capture various phases of partnership development.

A Phase Theory of Partnership Growth

Although the SEMIS Coalition was founded to strengthen teachers', students', and community members' abilities to re-vitalize commons-based practices, it did not start out with a ready-made and broadly shared vision and set of trusting relationships. Rather, it was through the course of partnership formation and working with communities and community partners, that our vision as a coalition became more inclusive (Lupinacci, 2013). This continues to be a messy process filled with stumbles and mistakes. Over time, and through reflection with our partner members on both our successes and challenges (Lowenstein, Frenzel & Schott, 2014), we began to identify qualities of partnership interactions and see patterns of these qualities that could be conceptually lumped into phases of partnership growth. We also began to notice that the partnerships that were able to remain at a "high" level were the ones that corresponded more and more with a holistic and eco-democratic way of being together in community. We realized that we needed a language to describe what we were experiencing in those partnerships. We began to ask ourselves, why do some of our partnerships fail, while others become transformational? In short, we added a dimension of time to our partnership orientation heuristic and zoomed in on the concrete qualities of interactions between organizations and individuals in those organizations. At the same time, we began to be influenced by ideas emerging in the field of psycho-social growth (Drago-Severson, 2009; Selman, 2003; Adaljarnardottir, 2010) within social contexts, and borrowed some language there that helped us to describe what we were seeing. For example, Eleanor Drago-Severson's constructive-developmental theory for adults includes a phase theory where adults move from instrumental to self-transformational ways of knowing. Instrumental knowers focus on following rules to benefit themselves. Self-transforming knowers welcome challenges to their own assumptions. We borrow Selman's (2003) language of unilateral, or "one-way," rule-based relationships, and reciprocal, "2-way," collaborative relationships to describe the first two phases of our model (p. 31).

Figure 2

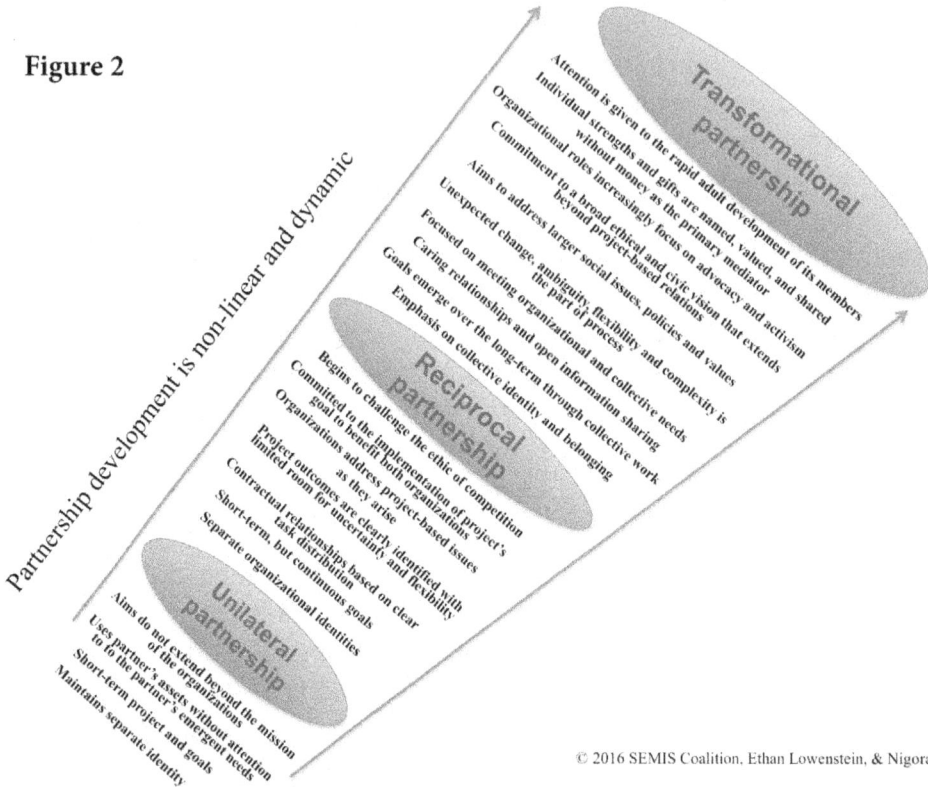

Partnership development is non-linear and dynamic

Transformational partnership

Attention is given to the rapid adult development of its members
Individual strengths and gifts are named, valued, and shared
Organizational roles increasingly focus on advocacy and activism
Commitment to a broad ethical and civic vision that extends
without money as the primary mediator
Aims to address larger social issues, policies and values
beyond project-based relations

Reciprocal partnership

Unexpected change, ambiguity, flexibility and complexity is
Focused on meeting organizational and collective needs
Caring relationships and open information sharing
Goals emerge over the long-term through collective work
Emphasis on collective identity and belonging
Begins to challenge the ethic of competition
Committed to the implementation of project's
goal to benefit both organizations
Organizations address project-based issues
as they arise
Project outcomes are clearly identified with
limited room for uncertainty and flexibility
Contractual relationships based on clear
task distribution

Unilateral partnership

Short-term, but continuous goals
Separate organizational identities
Aims do not extend beyond the mission
of the organizations
Uses partner's assets without attention
to the partner's emergent needs
Short-term project and goals
Maintains separate identity

The model (see Figure 2) describes three developmental phases of partnerships, moving from simple to complex, and from unilateral to transformational. The model can be used to more accurately articulate, and reflect on, an organization's theory of action in terms of partnership formation, as well as to analyze partnerships in action. Below we describe each phase with its descriptors. We couple the description of each phase with a vignette from the SEMIS Coalition experience.

Unilateral partnership.

At this level of partnership, each individual organization operates using a language of commodity exchange. It is aware of and driven by its own "needs," enters into partnerships when they are "profitable," knows what "resources" and "assets" other organizations have, and "mines" these "resources" to meet its own mission. In unilateral partnerships in education, the language of "programs," "projects," and "implementation" is used and exact "deliverables" are identified and quantified into "data." Partnerships are bound by "contracts" and "memos of understanding" that spell out short-term activities and goals and measure "success" by the "reaching" these goals.

"Progress" toward "goals" creates a culture of perfectionism and linear "development," not one of fallibility, journey, and mystery. In unilateral partnerships, there is no effort to build a common vision based on a broad sense of the public good, nor do organizations dedicate resources to relationship building activities that don't have short-term benefits or clearly defined "project outcomes." Coalitions built on unilateral partnerships never get off the ground because they are brittle and crack easily under pressure. Actions within these partnerships tend to be unethical because they are uncritical and unaware of the institutional power relationships and culture within which they function. This culture of "contract" rather than "covenant" is based on zero-sum competition that emphasizes an ethic on "efficiency," and "productivity" over mutual aid, affection, and gift giving. Anyone who works in a non-profit organization will recognize immediately that the language and rules that most non-profits, universities, governmental bureaucracies, foundations and government grant systems have in place reinforce unilateral organizational orientations. It is important to note that these cultural and institutional systems enact themselves in organizations regardless of these organizations' missions having opposite aims. This "double-bind" leads to a case where the cultural mindset of unilateral partnerships leads "good people" to do questionable or "bad things."

The SEMIS Coalition experience.

In reflection, when SEMIS began, we were still somewhat unaware of the cultural and institutional structures that bound us and the extent to which, even with the vision of cultural transformation of the founders, we still enacted unilateral partnerships with some of the partner-members of the Coalition. In retrospect, we were not yet, in fact, a "coalition." Contracts with member organizations were standard practice as were memos of understanding with inflexible resources and goals. In some instances, when time on the contract ran out, partner organizations ceased to come to Coalition events. Group fundraising was not a common practice, and in most cases we did not have the density of relationships, and a clearly established culture of covenant based on disciplines of neighborliness. As a Coalition we had not yet had time to form a common vision, nor articulate our partnership approach. In short, we did not yet have an advanced mental model and language through which to talk about partnerships even though we had an espoused philosophical orientation toward membership, collaboration, belonging, and affection. As a result, some organizations chose not to stick with the Coalition beyond this initial phase or became dormant for many years. Partners' roles remained unclear since a common vision and collective traditions to anchor this vision had not yet been created. Because unilateral cultural norms were dominant in some of our partnerships, conflicts arose with several organizations at this time. In some ways, this was a painful period, but we learned a lot, and were able to loop this learning back into modifying our organizational structures and creating traditions to anchor the culture we wanted to create (Lupinacci, 2013).

Near the end of our organizational infancy, and during the period when we started to articulate our mental models, and use these to reflect organizationally, a conflict arose at one of our partner schools. Because we were now beginning to have the language to describe it, we were able to successfully make informed decisions about this partnership. The school was a charter school, run by a for-profit management company. Originally, the school had a principal with a strong eco-democratic vision and who founded the school in collaboration with the Coalition of Essential Schools, an organization that emphasized teacher reflection and community building. When the principal left the school, several years into working with the SEMIS Coalition, the qualities of our partnership shifted dramatically and we felt a visceral sense of paradigm change. The new principal, as opposed to the predecessor, approached partnership with a unilateral mindset, and we became acutely aware of the cultural mismatch between this mindset and that of SEMIS, as a coalition.

The change of administration brought hierarchical rules for teachers at the school, and mandates to teach to the test and only engage in professional development activities that had this goal. The collaborative culture they had begun to develop at the school was systematically jettisoned, and Coalition of Essential schools group protocols were rarely used under the new administration. We became aware that the school viewed SEMIS as a "service-provider" and commodity instead of a co-created community. Because SEMIS offered high quality "professional development," the school saw us as an "intervention" that could be used to improve "teacher skills," and show "evidence" of "academic achievement," within a year of "program" "implementation." Unfortunately, the ultimate risk of state takeover in Michigan generally, and Detroit specifically, forces this unilateral mindset, and this school came to represent most schools in today's urban areas in the US.

Because we became more aware of what was happening at the school, and in our partnership, the SEMIS leadership team purposefully decided to explore whether we could work within this context to stretch this partnership to a reciprocal level and learn from this experience. We did not make this decision naively, and with our model as our reference point, we generated a legal contract that clearly stated the partnership "deliverables" that the school needed. We charged the school a fee commensurate with our efforts, because we needed funds. Within this context of each organization considering only their own needs, we wanted to determine whether we could collectively develop a more reciprocal relationship while operating within the heart of the dominant culture of schooling. Within a period of months, it became clear that our experiment was going to fail, in spite of our efforts to push the boundaries of the partnership. Because their goal was efficiency, and the new administration saw us as "service-providers" who conducted "in-service professional development," instead of seeing themselves as members of a strength-based coalition, they refused to dedicate organizational resources to relationship development and the formation of a common and broad ethical and civic vision. The channels of communication with the school became muddled and information we needed to be successful at the school was not shared with us. As communication dwindled, we also became less motivated to dedicate time and energy to this partnership when we had many other schools that needed our attention. We imagine that from their perspective, they also might have sensed that they were not going to get from us what they wanted, and saw the withdrawal of our attention to the school as a "breach of contract." Because of our partnership model, we rapidly assessed that this partnership was terminally unilateral and diplomatically worked with the school to end the contract, which effectively ended the relationship. We were able to leave on good terms, and we credit our ability to do this to our consciousness and transparency about what was occurring and why, and having an accurate language to describe partnership dynamics. It is important to note that being in a unilateral mindset and having institutional structures does not mean that the school's leaders did not care for the children or want the best for them. It does mean, however, that the school's leaders did not have a partnership orientation that was eco-democratic, and therefore could not facilitate underlying organizational and broader cultural transformation toward commons-based values and practices.

Reciprocal partnership.

At this level of partnership, each organization is aware of each other's needs and works toward understanding and supporting its own mission and interests as well as the mission and interests of partners. Over time, identities in reciprocal partnerships can begin to move from "partnership" to "coalition," because a genuine concern about the welfare of the partnering organization is present. At this phase, partnerships are still project-based and do not intend to stretch beyond agreed upon "contract" details. However, in reciprocal partnerships there is a level of mutual trust, a growing expectation that current projects will lead to future projects, and there is a willingness to discuss difficult partnership issues as they arise during collaborative activities. Relationships tend to be cooperative and use a language of "task distribution." Trusting relationships can be accelerated in this phase when organizations are purposeful in using appreciative and strength-based language in determining organizational roles within projects, as well as being honest about the limitations of their own organization and its capacity. In the reciprocal partnership phase, organizations may keep each other's missions and perspectives in mind, and be respectful and caring in their orientation to a partner's mission, but not yet enter into collective reflection about how they might co-develop a common culture based on a broad civic and ethical vision that stretches beyond each organization's necessarily limited mission. Because reciprocal relationships operate on the border between unilateral and transformational relationships, they can become strained when under pressure. For example, task distribution and the time bound nature of contractual relationships can lead to a lack of flexibility when the context of projects changes, which often happens in urban school environments, or when personnel leave an organization unexpectedly because contingent work turns into more stable work or because individuals face difficult life circumstances like illness. Under such conditions, people in partnering organizations may become resentful as they take on unanticipated tasks without changing partnership goals or processes. Being in the reciprocal phase of a partnership is not a pure state, and individuals may begin to enter into relationships of vision co-development across organizations, but not yet be part of relational networks within and between organizations that are dense enough to cause a full organizational shift in its approach to partnership. Although the possibility of coalition is present in reciprocal partnerships, organizations are prone to reverting back to a unilateral mode if formal organizational leadership changes, which often happens in today's non-profit and school world, or if language and structures are so rigid that there is no flexible time dedicated to relationship building and organizational transformation outside of "funded" activities. Partnerships in the SEMIS Coalition have sometimes gone dormant for many years because of these conditions, only to emerge when the possibility of striving toward transformational partnerships re-surfaces with personnel changes or funding that provides organizational stability and flexibility. In cases like the one described below, struggling through reciprocal relationships, and reflecting on this struggle can help partners identify what needs to be conserved in their practices and what needs to be transformed in order to further move toward a coalition based on covenantal and eco-democratic ethos and practices.

The SEMIS Coalition experience.

One of the long-time members of the SEMIS Coalition is an arts-based organization that, since the beginning of its partnership with the Coalition, has moved between reciprocal and transformational phases within the Coalition. The former Director of this organization is a visionary leader with a strong and broad eco-democratic vision for the role that the arts and inter-generational dialogue can play as commons-based cultural practices. In the early years of the Coalition, she was one of our first partners to co-facilitate parts of the SEMIS Summer Institute without monetary compensation. This is now a regular and common practice of ours.

Even with the contributions that the Director of the partner organization made to the SEMIS Coalition both as an individual, and in the eco-democratic vision that she shared with SEMIS' founders, our two organizations were not yet in a transformational phase of our relationship as organizations. We did not have a dense network of relationships between the two organizations, and organizationally, we had no knowledge of each other's inner workings and culture. So, when we applied for a fully collaborative grant—one that would require our staff to work with each other closely in chaotic school environments and co-develop professional learning seminars for our teachers, we encountered many challenges. The original grant had some budget flexibility, but when we received it, the grantor told us that we would only receive two-thirds of the funding while being held to the same outcomes. This forced us to distribute tasks quickly, and we didn't have the time to collectively think through what we were willing to give up in our processes and activities. Staff in both organizations was part time during the grant period. SEMIS Coalition staff, however, had much more flexibility because we were university-based, and could rearrange resources if needed. The result was that we found ourselves wanting to be in transformational relationship with each other, but in practice were still located in structures and conditions that led us to act within a reciprocal phase of partnership. Both organizations had an idealistic orientation, because even though the Director of the partner organization and SEMIS staff shared a common vision, we did not have the dense relational network of staff necessary to reflect in a sustained way on organizational structures and conditions and manage the differences between our organizations. For example, going into the collaborative project, we did not have knowledge of how each organization tracked and allocated staff time, the protocols it used for documenting the work of staff, norms for communication, or traditions of planning with teachers. Our organizational mismatches and contractual agreements, under difficult work conditions, put an intense amount of pressure on the partnership. The perfect storm came when the arts' organization's Chief Financial Officer left unexpectedly, forcing the organization's leadership to pick up the slack, and the SEMIS Co-Director, who was responsible for many of the grant activities, had a serious car accident and did not have the ability to engage as the contract dictated without a personal toll on her health.

What saved our partnership was its reciprocal nature and its goal of moving toward a transformational partnership. A unilateral partnership would have disintegrated within a matter of months. However, in our case, each partner communicated with the other openly about its struggles and actively attempted to take the other organization's perspective. Even though day-to-day project operations were a struggle throughout the year, there was an overall feeling of mutual trust between organizations, and we tried as often as possible to use strength-based and appreciative language to talk about each other's strengths and the real limits of each other's organizational capacities. Because day-to-day communications were sometimes challenging, for example, we had to make sure that we did not revert to a unilateral way of thinking when staff between the organizations disagreed or when incompatible elements of organizational cultures and protocols clashed. At the time, we did not yet have our partnership models fully articulated, so we did not have the theory or language to identify or fully explain what was happening. Both organizations knew we were in a nightmare project scenario, but because our partnership was not completely time-bound, and we expected further collaborative projects, we were ultimately able to put the particular project in perspective. It was, however, a very painful year for both organizations, and keeping our beliefs and interactions in the reciprocal phase took tremendous effort. It was no surprise that after this particular project was finished, our partnership went dormant. Because of some shared vision and the long history of our partnership, however, there continued to be potential for transformational partnership, and recently our organizations have begun working closely again with renewed energy and understanding.

Transformational partnership.

We use the word "transformational" in describing this phase of partnership to denote three primary qualities of this phase. The first is that partners who enter this phase are engaging in a cultural break away from consumer culture and toward a commons-based ethos and practice. The second is that partners in this phase openly invite challenges to the primary assumptions that undergird their organizations' structures and processes. Finally, partners in this phase are committed to developing the habits of mind, heart, and hands to enact the first two qualities, and see this necessarily as a community endeavor. At this phase of partnership, each partner organization supports the other in the service of broad civic and ethical purposes that encompass, and go beyond, the specific mission and practice of each organization. Organizations are open to change, comfortable with complexity, are process oriented, and are grounded in cultural norms and practice that emphasize belonging, trust, and affection. Because the purpose is moving toward a common civic and ethical vision, partners in this phase do not "measure" their success by the "progress" they are making in a particular project, but rather whether their collaborative activities enact a process that will conserve existing, and/or bring about, new eco-democratic values and cultural traditions in their organizations and in the coalition in which both the partners are members. Here, partners dedicate a lot of effort to looking at patterns of growth, sensing what activities want to emerge given current conditions, and determining when and where there is sufficient collaborative energy to push the collaboration deeper and broader. Partners are cognizant of the fact that cultural change takes decades, and that their progress toward their vision will necessarily be non-linear and chaotic and will sometimes appear on the surface to take one step forward and two steps back. Partners in this phase operate on a "community" sense of time and development of "slow knowledge" (Orr, 2002), as opposed to a mechanistic industrial sense of time based on consumerism. What we describe here is not new, but has been practiced throughout human history by many indigenous and land based cultural traditions (Esteva & Prakash, 1998; Mander & Tauli-Corpuz, 2006). In the transformational phase, processes of resource exchange operate under an ethic of gift-giving or barter. Contracts and memos of understanding are no longer used, or are only used as a purposeful strategy for maneuvering within power structures, not as the defining rules of the relationship. For example, an organization might have a contract with another, but both organizations realize this is just playing a game for an outside audience—there is no illusion that this transaction or the language of the contract represents their relationship, nor does the contract influence their decision-making within it. Said another way, partners in the transformational phase must both speak a language and use metaphors that those with institutional power understand, at the same time that they work to establish a parallel mutualist culture. They must also be able to meta-cognitively take perspective on their code-switching. Organizations in the transformational phase, when they approach a true cultural break, can look "crazy" and "touchy-feely" to those not in this phase of understanding. The emotionally-laden and affectionate language that they often use, for example, can sound weird at best, and cultish at worst, to those squarely positioned within contractual culture. This phase requires that organizational leaders are systems thinkers and that there is a density of individuals in the organization that are at high levels of adult development. For this reason, organizations with transformational and eco-democratic orientations dedicate a lot of resources to the sustained psycho-social growth of their members and to collective discourse and root cause cultural analysis.

Members of organizations in this phase consistently use strength-based and appreciative language, as well as publicly value each other's eccentricities. In SEMIS, for example, we often tell new members that our gatherings are more like a "family reunion" than a "conference" or "professional development (PD)." We also talk about members' contributions to "the commons," and speak in terms of responsibility to the SEMIS Coalition "community," rather than the "rights" and "obligations" of membership.

Ironically, the ultimate results of being in this phase, although it may take years to get there, are attributes valued by market-driven cultures such as incredible amounts of flexibility, resilience, and efficiency of collaborative efforts. For example, in the transformational phase, individual organizations contribute to the common pot of resources when they get them, and don't when they don't get them. Currently in SEMIS, when a member needs help, they can call on people from dozens of organizations with different strengths and knowledge, and usually get an instant and engaged response from them. That member understands that the same will be expected of them when they are called upon to share their own strengths with other members. They also understand that there are times when a member needs to take a break from the relationship based on other demands on their energy and time. One of our successes in SEMIS has been the creation of a true culture of abundance. The creation of such culture does not come without growing pains.

The SEMIS Coalition experience.

We do not want to give readers the sense that the SEMIS Coalition sprang from thin air upon its founding. Instead, it was deeply rooted in the relationships that its founders and different members of the Coalition's Steering Committee brought to the organization over time. Some of these relationships had been there for over a decade. And these relationships, in turn, sat within visionary community efforts spanning over 40 years. One example, is that of the relationship between members of the SEMIS Steering Committee and the James and Grace Lee Boggs Center to Nourish Community Leadership. Members of the SEMIS Steering Committee and other members of the Coalition had been in relationship with Grace Lee Boggs, as well as a group of educators and community members who gathered in the early 2000's to have Freedom Schooling discussions. This essay's lead author, Ethan Lowenstein, for example, had a continuous relationship with several members of the original Freedom Schooling group and had been collaborating closely with Amanda Rosman, a teacher in this group, in an almost continuous fashion since 2002 (Lowenstein, 2010; Lowenstein, 2016). A couple of years after Ethan joined SEMIS, Amanda came to a SEMIS gathering or two and then asked for help with a new and ambitious project. Amanda, along with Julia Putnam, and Marisol Teachworth, two long-time members of the Boggs Center and Freedom Schooling discussions, were in the process of creating a school founded on the ideas of Grace Lee and Jimmy Boggs, and the idea of Place-based Education (Putnam, 2011). The initial step in this relationship was SEMIS assistance with creating a curricular foundation for the school. The school happened to have money resources from a planning grant. They gifted these resources to SEMIS, as SEMIS needed them. SEMIS gifted its knowledge because the Boggs School needed it. Both organizations freely gave these gifts out of love, and in service of creating healthy and just communities and schools. Although there was a written contract and money "exchanged" for "services," there was never an illusion that its rules represented the relationship; we never counted the hours we "put in," nor did we ever use the language of exchange and service internally. As needs of each partner emerged, they were sensed and attended to with care and respect. A critical one was our ability in the SEMIS Steering Committee, because we had high levels of trust with Eastern Michigan University (EMU) administration, to help the Boggs School get its charter from EMU at a point when finding an authorizer who understood the Boggs School's vision was challenging and morale was very low.

From a partnership perspective, this is an example in which members of both organizations shared a broad civic and ethical vision from the outset, but did not yet have the density of relationships between organization members and the formal inter-organizational structures needed to enact a fully eco-democratic relationship within a new and complex context. For instance, starting a school is complicated business, especially when it attempts to walk the walk of enacting covenantal values while having to follow an astonishing number of hierarchical rules and provide "data" to many "stakeholders" to survive. Federal, state, and city school system policies change year to year, and this creates an unpredictable and dynamic context where just when you think you have something in place, the ground shifts under your feet. The SEMIS Coalition is also a very complicated organization with its own set of traditions and aims, and its own short amount of time to grow to an extent that we can attract the funding to sustain our efforts. Both organizations were, and continue to be, under a lot of pressure!

Taking our partnership from one that was somewhat idealistic, given the complexity of the new context, to one that at moments has been holistically eco-democratic, has taken five years and has at times been a very painful process because our energies have been pulled in many different directions and we have sometimes been unable, as individual organizations, to meet each other's needs and fully care for each other. Having dense relationships and a community mindset is easily romanticized when in fact it is double-edged—as ones' relationships become deeper and more numerous, one finds oneself ethically responsible for many and it becomes literally impossible to meet the needs of all of one's partners at any one point in time. For example, SEMIS staff dedicated a lot of focused energy to supporting the Boggs School. In reciprocal relationships, efforts are exchanged, and defined by specific projects in a relatively short window of time. It was a painful process to recognize that this belief system still influenced our thinking, and also that it was going to be impossible to fully meet each other's needs over a short period of time, even when these needs were great. Members of the Boggs School experienced feelings of guilt because during the first couple of years of the school, they could not participate in SEMIS and give to Coalition efforts in an "equal" way that SEMIS had given to them. SEMIS staff also experienced sadness that the Boggs School, in its first years, was not wholly active in Coalition activities because of its other demands, and had a fear that the relationship was weakening. It took many discussions, sometime involving tears, to work through these feelings of guilt on both sides and understand how our feelings and expectations were being informed by the cultural mindsets around partnership we had inherited. Ultimately, the mental model of partnership that we were formalizing, which was informed by commons-based cultures we were a part of, helped us navigate our partnership in healthy ways. The mental model allowed us to talk about the nature of transformative partnerships, the long-view of time, being flexible and process driven, and developing a culture of gift-giving and patience for mistakes and failure. Most recently, we have collectively reflected in the Coalition as a whole and in our partnership specifically about guilt and shame, and how to create traditions and routines in our organizations and in the Coalition for collectively working with these feelings in deep and sustained ways.

Concluding Thoughts

In this short essay, we have attempted to introduce two mental models for thinking about partnerships and ground these models in experiences from the SEMIS Coalition. We have found through our own experience, and working with many other organizations, that the formation of transformational partnerships that are in line with a broad ethical and civic vision is no simple affair. Organizations can move back and forth between orientations and phases of partnership depending on a complex set of circumstances. Organizations will also likely have partnerships in different phases of development that require different approaches. Although partnerships are complex, they are not, however, unique. Partnerships have patterns, and by articulating the qualities of these patterns, organizations can more effectively navigate through them. We hope that the reader has gotten a sense of the messy, imperfect, and dynamic nature of partnerships that attempt to enact eco-democratic values within school and organizational systems dominated by hierarchical belief systems and structures. The ideas in this essay are not new, and in many cases, can be traced back millennia. We sense that we are in an historic moment when people in many organizational contexts, from business to educational leadership, are thinking about similar issues and trying to work through the cultural contradictions and possibilities of our time.

We look forward to learning and growing with readers whose experience and vision resonates with our own reflections, and hope this essay has been helpful in helping them to reflect on their own efforts in their particular place.

❖ ❖ ❖

Ethan Lowenstein, Ph.D. is a Professor of Curriculum and Instruction at Eastern Michigan University and the Director of the Southeast Michigan Stewardship Coalition (semiscoalition.org) a professional development network of schools, community partners, K-12 teachers, and university educators working together to address serious ecological and social problems in Southeast Michigan. Dr. Lowenstein has close to two decades of experience in facilitating and researching teacher professional development in moral, civic, and place-based education.

Nigora Erkaeva, ABD, is a doctoral candidate at Eastern Michigan University and currently working on her dissertation. She also teaches an undergraduate social foundations of education course that integrates a place-based approach. Her research interests include ecojustice education, place-based learning, social justice, postcolonial theory and the role of education in addressing ecological and social injustices in our communities.

References

Adalbjarnardottir, S. (2010). Passion and purpose: Teacher professional development and student social and civic growth. In T. Lovat, R. Toomey & N. Clement (Eds.), *International research handbook on values education and student wellbeing* (737-763). London, UK: Springer.

Barnard, C. I. (1968). *The functions of the executive.* Cambridge, UK: Harvard University Press.

Block, P. (2009). *Community: The structure of belonging.* San Francisco, CA: Berrett-Koehler Publishers, Inc.

Block, P., Brueggemann, W., & McKnight, J. (2016). *An other kingdom: Departing the consumer culture.* Hoboken, NJ: John Wiley & Sons, Inc.

Bowers, C. A. (2001). *Educating for eco-justice and community.* Athens, GA: University of Georgia Press.

Bowers, C.A. (2005). *The false promises of constructivist theories of learning.* New York: Peter Lang.

Bowers, C. A. (2006). *Revitalizing the commons: Cultural and educational sites of resistance and affirmation.* Lanham, MD: Lexington Books.

Bowers, C. A., & Flinders, D. J. (1990). *Responsive teaching: An ecological approach to classroom patterns of language, culture, and thought.* (Vol. 4). New York, NY: Teachers College Press.

Bowers, C. A., & Martusewicz, R. (2006). Revitalizing the commons of the African-American communities in Detroit. In C. A. Bowers (Ed.), *Revitalizing the commons: cultural and educational sites of resistance and affirmation* (47-84). Lanham, MD: Lexington Books.

Bringle, R.G., Clayton, P.H., & Price, M.F. (2009). Partnership in civic service learning and civic engagement. *Partnerships: A Journal of Service Learning and Civic Engagement,* 1(1), 1-20.

Demarest, A. B. (2014). *Place-based curriculum design: Exceeding standards through local investigations.* New York, NY: Routledge.

Dowling, B., Powell, M. & Glendinning, C. (2004). Conceptualizing successful partnerships. *Health and Social Care in the Community,* 12(4), 309-317.

Dorado, S., & Dwight, E. G. (2004). Service-learning partnerships: Paths of engagement. *Michigan Journal of Community Service Learning,* 11(1), 25-37.

Drago-Severson, E. (2009). *Leading adult learning: Supporting adult development in our schools.* Thousand Oaks, CA: Corwin Press.

Enos, S., & Morton, K. (2003). Developing a theory and practice of campus-community partnership. In B. Jacoby & Associates (Eds.), *Building partnerships for service-learning* (20-41). San Francisco, CA: Jossey-Bass.

Esteva. G., & Prakash. M. S. (1998). *Grassroots Post-modernism: Remaking the Soil of Cultures.* New York, NY: Zed Books.

Foucault, M. (1972). *The archaeology of knowledge.* New York, NY: Pantheon Books.

Gruenewald, D. A. (2003). The Best of Both Worlds: A Critical Pedagogy of Place. Educational Researcher, 32(4), 3-11.

Greenwood, D. (2013). A critical theory of place-conscious education. In M. Brody & R. B. Stevenson (Eds.), *International handbook of research on environmental education* (93-100). New York, NY: Routledge.

Greenwood, D. (2016) Reclaiming the promise of place: An interview with David Greenwood. Retrieved from: https://www.bankstreet.edu/occasional-paper-series/33/reclaiming-the-promise-of-place/

Gulamhussein, A. (2013). Teaching the Teachers: Effective professional development in the era of high stakes accountability. Retrieved from The Center for Public Education web site: http://www.centerforpubliceducation.org/Main-Menu/Staffingstudents/Teaching-the-Teachers-Effective-Professional-Development-in-an-Era-of-High-Stakes-Accountability/Teaching-the-Teachers-Full-Report.pdf

Hoppe, B. & Reinelt, C. (2010). Social network analysis and the evaluation of leadership networks. *The Leadership Quarterly,* 21, 600-619.

Ignaczak, N. (2014, September 30). Place-based education motivates students by connecting learning to community. Model D. Retrieved from http://www.modeldmedia.com/features/place-based-education.aspx

Lipmanowicz, H. & McCandless, K. (2016). Retrieved from http://www.liberatingstructures.com

Lowenstein, E., Martusewicz, R., & Voelker, L. (2010). Developing teachers' capacity for ecojustice education and community-based learning. *Teacher Education Quarterly,* 37(4), 99-118.

Lowenstein, E. (2010). Navigating teaching tensions for civic learning. *Learning and Teaching (LATISS),* 3(1), 32-50.

Lowenstein, E., Frenzel, J., & Schott, J. (2014). Creating Transformative Partnerships for Place-Based Education: Tales from the Southeast Michigan Stewardship Coalition. Paper presented at the Third Annual EcoJustice and Activism Conference 2014, Ypsilanti, MI. Retrieved from https://docs.google.com/file/d/0B8CDnY9sPS7BQmdmU1VZckFwb2c/edit

Lowenstein, E. (2015). Green living: Growing visionary educational communities. Retrieved from: http://www.crazywisdomjournal.com/featuredstories/2015/12/31/green-living-growing-visionary-educational-communities

Lupinacci, J. (2013). The Southeast Michigan Stewardship Coalition: A Deep Design of Eco-Democratic Reform that is Situational, Local, and In Support of Living Systems (Doctoral dissertation, Eastern Michigan University). Retrieved from http://search.proquest.com.ezproxy.emich.edu/pqdtft/docview/1450044555/E5C87C550EAA460DPQ/1?accountid=10650

Mander, J., & Tauli-Corpuz, V. (Eds). (2006). *Paradigm wars: Indigenous people's resistance to globalization.* San Francisco, CA: Sierra Club Books.

Martusewicz, R. A., Edmundson, J., & Lupinacci, J. (2015). *EcoJustice education: Towards diverse, democratic, and sustainable communities* (2ⁿᵈ ed.). New York, NY: Routledge.

Macdonald, S. & Chrisp, T. (2005). Acknowledging the purpose of partnership. *Journal of Business Ethics,* 59 (4), 307-317.

Plumwood, V. (1997). *Feminism and the mastery of nature.* New York, NY: Routledge.

Putnam, R. D. (2001). *Bowling alone: The collapse and revival of American community.* New York, NY: Simon and Schuster Paperbacks.

Putnam, J. P. (2011). Another education is happening. *Monthly Review,* 63(3), 56.

Senge, P. M., Smith, B., Kruschwitz, N., Laur, J., & Schley, S. (2008). *The necessary revolution: How individuals and organizations are working together to create a sustainable world.* New York, NY: Crown Publishing Group.

Senge, P. M., Scharmer, C. O., Jaworski, J., & Flowers, B. S. (2005). *Presence: An exploration of profound change in people, organizations, and society.* New York, NY: Crown Publishing Group.

Scharmer, C. O., & Kaufer, K. (2007). *Leading from the emerging future: From ego-system to eco-system economies.* San Francisco, CA: Berrett-Koehler Publishers.

Selman, R. L. (2003). *The promotion of social awareness: Powerful lessons from the partnership of developmental theory and classroom practice.* New York, NY: Russell Sage Foundation.

Smith, G. A. (2002). Place-based education: Learning to be where we are. *Phi Delta Kappan,* 83(8), 584-594.

Smith, G. A., & Sobel, D. (2010). *Place-and community-based education in schools.* New York, NY: Routledge.

Sobel. D. (2005). *Place-based education: Connecting classrooms and communities.* Great Barrington, MA: The Orion Society.

Wheatley, M. J. (2007). *Finding our way: Leadership for an uncertain time.* San Francisco, CA: Berrett-Koehler Publishers.

FREEDOM, JUSTICE AND SUSTAINABILITY:
DO WE REALLY KNOW WHAT WE ARE DOING?

Rolf Jucker

"We don't have the courage nor the capacity to admit that meaning for our individual and collective lives cannot be provided anymore by a religion or an ideology, cannot be given to us as a gift; that we have to create it ourselves." (Castoriadis 2005: 327; my translation)

Eco-justice: vision and current unsustainable reality

Max Horkheimer and Theodor W. Adorno have defined the state of human liberation as such: freedom from oppression within oneself, freedom from oppression through other people, absence of exploitation of nature (1986: 61). You could also describe this as a situation where there are no hierarchies between people, where there is no abuse of power, or indeed no power structures that allow such an abuse. In such a world there would be no exploitation of nature or other species or other people elsewhere in the world or future generations to facilitate your life-style. In other words, there is no way that this life-style would be beyond what a just global distribution of ecological footprints would allow. But Horkheimer and Adorno's vision importantly also focuses on the world within: in this vision there is also no oppression from within: no belief systems or traditions or social structures or peer or family pressure to force us into an acceptance of subjugation which undermines self-determination, free will, freedom from fear and the true development of our human potential. To me this is the vision of a truly human society which in my understanding is also the vision of the eco-justice movement. It is a vision with a long history of millions of people fighting for it since many decades, even hundreds and thousands of years (see for example Zinn 1996). It is the vision of becoming truly human, without the shackles of slavery, religion, wealth, aristocracy, economic exploitation, capitalism, communism, nationalism, patriarchy, sexism, ... It is, in short, the vision of the enlightenment which Immanuel Kant has so aptly captured in the following words in 1784:

> "Enlightenment is man's emergence from his self-imposed immaturity. Immaturity is the inability to use one's own reasoning without guidance from others. Self-incurred is this immaturity when its cause lies not in lack of reasoning, but rather of resolve and courage to use it without direction from others. Sapere Aude! Have courage to use your own mind! Thus is the motto of Enlightenment." (Kant 1784)

With John Lennon's "Imagine" in the ear (http://www.lyrics.com/imagine-lyrics-john-lennon.html, accessed 28 March 2016), it seems easy to imagine such a world where the aims of the French Revolution become reality: "liberty, equality, fraternity"; a world where a person is a human being and not a refugee, where woman and girls are equals in a true sense with men and boys, and not pressed into a state of dependency through moral laws which have long lost their validity; where all people respect the fundamental values of an open, democratic, secular society, based on knowledge and understanding, and not on myths, oppressive belief systems, autocratic rules and power structures based on status and wealth.

But, of course, we all know how far off we are from a reality which at least starts in some ways to resemble this vision. I note but four of the most obvious issues:

- *capitalism:* Why do we stick to an economic system which consistently destroys democracy and sound social ties, in addition to the planet? We know that our way of doing business, in fact, our idea that business and the economy are the core of our lives is not just plain wrong, but in fact the most destructive force ever unleashed on nature and human beings. Just read Sven Beckert's *Empire of Cotton. A Global History* (2014) or David Graeber's *Debt. The first 500 years* (2011) as impressive illustrations of this.

- *power:* As Noam Chomsky said a long time ago: "power never self-destructs" (1994). Yet are we really challenging the absolutely supreme power of the political and economic elites worldwide? Contrary to popular belief, the internet has not only given the people more self-determination and power. It has also amassed a concentration of power and knowledge in the hands of a few, not democratically controlled corporations (Microsoft, Apple, Facebook, Google, Amazon), and they, together with their state-brethren in the NSA, control us in a way that George Orwell's *1984* (1949) or Aldous Huxley's *Brave New World* (1989) look just like silly kindergarten games in comparison (see the film *Citizenfour* on Edward Snowdon for this: https://citizenfourfilm.com/, accessed 25 March 2016). But we are so engrossed in our individual happiness that there is no public discourse about the absence of democratic self-determination or, indeed, what a truly democratic political system might look like (see Lummis 1996).

- *wealth:* I am convinced that we need to make the rich a lot poorer to narrow the gap between rich and poor. In their thoroughly researched book *Spirit Level* Wilkinson & Pickett show that equal societies consistently score significantly better than unequal ones. Almost all problems which turn our modern societies into "social failures" are more common in unequal societies: "level of trust, mental illness (including drug and alcohol addiction), life expectancy and infant mortality, obesity, children's educational performance, teenage births, homicides, imprisonment rates, social mobility" (Wilkinson & Pickett 2010: 18–19). The conclusion from their book: "The evidence shows that reducing inequality is the best way of improving the quality of the social environment, and so the real quality of life, for all of us. (...) this includes the better-off." (ibid. 29)

- *destruction of our life-support system planet earth:* Let us be honest here: we have dangerously transgressed already three of the seven critically important planetary boundaries (Rockström *et al.* 2009), we are depleting crucial materials at an alarming rate (New Scientist 2013), twenty-four years after Rio the world is a dirtier (WCA 2014) and more systemically unsustainable place than ever before (LPR 2012: 9, Worldwatch 2013a and 2013b, eia 2013, Wardsauto 2015, NYT 2013, ScienceDaily 2012).

So I guess we are fairly clear about the vision or future we are prepared to fight for and we are fairly clear about the enemies which we need to fight in this process. I love the following quote from Chris Hedges because he takes no punches and doesn't flinch away from the uncomfortable fact that we have a fight, quite possibly a revolution, at our hands if we want to move from our unsustainable current world into a human world as outlined above:

> "We can cut our consumption of fossil fuels. We can use less water. We can banish plastic bags. We can install compact fluorescent light bulbs. We can compost in our backyard. But unless we dismantle the corporate state, all those actions will be just as ineffective as the Ghost Dance shirts donned by native American warriors to protect themselves from the bullets of white solders at Wounded Knee. (...) The oil and natural gas industry, the coal industry, arms and weapons manufacturers, industrial farms, deforestation industries, the automotive industry, and chemical plants will not willingly accept their own extinction. They are indifferent to the looming human catastrophe. We will not significantly reduce carbon emissions by drying our laundry in the backyard and naively trusting the power elite. The corporations will continue to

cannibalize the planet for the sake of money. They must be halted by organized and militant forms of resistance." (Hedges 2010: 293)

I am convinced that it is not so difficult to understand these issues. It all boils down to the power structures we have created within and between us as well as the exploitative systems of abusing nature to fuel our greed. Horkheimer and Adorno understood this well after fascism and the holocaust shone the spotlight so clearly on these issues in their times.

It is more difficult to pinpoint why there is so little progress towards an eco-just society anywhere in the world. This leads me to the main focus of this essay: I argue that we need some serious self-critical reflection on our own concepts and actions. As I have done above, it is very easy to blame others, the capitalists, the media, the internet, the corporate vandals and religious fanatics of this world. And I am even conceding that this blame is, of course, in most cases more than justified—as Chris Hedges argues.

Understanding the world with intuition, opinion and emotion?

But I am also increasingly convinced that part of the problem why we are not making any progress towards eco-justice, equality and democratic self-determination lies in the fact that we ourselves, in our efforts to further eco-justice, are hampered by what Leiva has so aptly termed "arrogance of ignorance" (2012). The past thirty years of intellectual discourse of postmodernism, cultural relativism and infatuation with all sorts of esoteric and traditional or indigenous 'knowledge' have led us to a state where we have lost much of our bearings and cannot easily distinguish anymore between knowledge and opinion, between truth and claim (see Jucker 2014). I think Slavoj Žižek is making a really important point here:

> "There are not just different forms of knowledge—the scientific, the magic, the social knowledge, etc. No, there is true and false knowledge. (…) We have to re-learn to argue in a tough way—even if this means that we are hurting people's feelings. Their concern, their pain is no measure for the truth. And truth, after all, should be our guide. Only then will we arrive at a universalism which will move human kind forward." (Žižek 2016; my translation)

Today, the arrogance of ignorance ("I am entitled to my own opinion, however stupid and factually wrong it may be") reigns supreme and usually cannot be challenged. Just think of the utter nonsense many of us are happy to believe in the area of food and health, veganism arguably being the most popular and most ideologically driven example at the moment: it is almost impossible to get so many fundamentals about life and death and systemic interdependence of species so wrong (just read Lierre Keith's *The Vegetarian Myth*, 2009).

But there is much more when it comes to the level of ignorance we still display when trying to make sense of the world. Much of it stems from our easy acceptance, induced by the above mentioned postmodernist, relativist and subjectivist trends, that there is no objective reality. Alan Sokal beautifully showed how much many of us have lost the ability to distinguish facts from fiction, sense from nonsense in his wonderful live-experiment with the renowned cultural studies journal *Social Text*. In the article where he testified to his scam, he writes:

> "What concerns me is the proliferation, not just of nonsense and sloppy thinking *per se*, but of a particular kind of nonsense and sloppy thinking: one that denies the existence of objective realities, or (when challenged) admits their existence but downplays their practical relevance. (...) Intellectually, the problem with such doctrines is that they are false (when not simply meaningless). There *is* a real world; its properties are *not* merely social constructions; facts and evidence *do* matter. What sane person would contend otherwise? Theorizing about "the social construction of reality" won't help us find an effective treatment for AIDS or devise strategies for preventing global warming. Nor can we combat false ideas in history, sociology, economics and politics if we reject the notions of truth and falsity." (Sokal 1996: 63; see also in more detail Sokal & Bricmont 1998)

My first example of such sloppy thinking is that we still trust in 'common sense' or 'intuition' when we should have learnt, historically and scientifically, that these are the things least likely to make us understand what really is going on. Most of the important advances in science in the last hundred years, for example in quantum physics—the one theory with the most accurate predictive quality humankind has ever developed –, have shown that intuition is a very poor guide to understanding the world. Just think of Einstein's theory of relativity (which, incidentally, has nothing whatever to do with postmodernist relativism mentioned above) which showed us that our 'intuitive', everyday assumption that mass, time and gravity is constant, is far from true; or our notion that matter is solid when it rather is empty space; or the fact that what we see with our eyes when we gaze into space is rendering only a very limited picture of what is really out there since there are a number of well-explained phenomena which make it impossible that our human vision 'sees' all there is (Crockett 2016).

In many indigenous cultures and particularly in eastern cultures—with an incredible allure to western people up to the present day—intuition is linked to the 'inner self', the 'essence' of us as a person, 'soul', the 'inner master' or whatever you might want to call it. The suggestion is that this 'inner core' can intuitively and more reliably than anything else know the truth about us, our feelings and the world. But this is misleading on various levels. Firstly, the notion of an 'autonomous', individual unit such as 'I' or 'self' is a culturally constructed narrative which has no substance in biological reality, as Donna Haraway has so aptly described:

> "I love the fact that human genomes can be found in only about 10 percent of all the cells that occupy the mundane space I call my body; the other 90 percent of the cells are filled with the genomes of bacteria, fungi, protists, and such, some of which play in a symphony necessary to my being alive at all, and some of which are hitching a ride and doing the rest of me, of us, no harm. I am vastly outnumbered by my tiny companions; better put, I become an adult human being in company with these tiny messmates. To be one is always to *become with* many." (Haraway 2008: 3-4)

But the notion of a coherent, inner, essential 'I' or self, which can be pinpointed, is also incorrect in terms of how the brain and consciousness works, as Francisco Varela explains (1999):

> "There are the different functions and components that combine and together produce a transient, nonlocalizable, relationally formed self … we will never discover a neuron, a soul, or some core essence that constitutes the emergent self of Francisco Varela or some other person." (quoted in Capra/Luisi 2014: 181)

We are truly relational beings, not isolated individuals:

> "Our existence is posited on our continued dialectic with the natural and social world that surrounds us, for as persons we cannot be monads, autonomous isolated individuals. I argue that our mental processes, and indeed consciousness, are created in and constituted by those relationships." (Rose 2006: 310)

This is rather important since the (false) construct of an 'autonomous individual self', independent of the life-support systems earth and social interaction, is still informing most of what we think and do, in eco-justice and elsewhere, in eastern, western and other cultures alike.

A second idea I would like to question is that there are these 'good' emotions versus the 'bad' mind, particularly popular in Eastern philosophy and religions like Buddhism. The slogan is, also in western 'How do I better myself' literature: "Less thinking. More feeling." (Brown 2013: 6) In yoga and meditation, the mind is constantly vilified as the great distractor, yet in fact we are nothing, literally and simply not alive without our brain/mind, which is always embodied, never transcendent (Capra/Luisi 2014: 142, 274). It is, after all, not a coincidence that the measure to determine whether somebody is alive or dead is to check if the brain is still working. There is no emotion, not gut-feeling without our brain processing the sensory information and actually producing an emotion, quite apart from the fact that emotions are heavily culturally co-determined and therefore hardly ever only personal. Or as Kristen A. Lindquist put it: "It goes without saying that the brain produces emotions—in this day and age, you'd have to be a pretty staunch dualist to argue otherwise. The big question that remains concerns how the brain creates emotions." (2016)

Thirdly, let me bring up another issue: overpopulation. Why do we pretend that the P (for population) in Ehrlich's "I = PAT" formula doesn't count? Paul Ehrlich's "I = PAT" formula is used to describe the impact of human activity on the environment: human impact (I) increases with increasing population (P), with increasing affluence (A) and with increasing resource intensity of technology (T) (Ehrlich 2013). Here, the 'arrogance of ignorance' quite literally is visible by fact that we are ignoring the entire issue for fear of being politically incorrect. In our societies it is a total taboo to talk about overpopulation, about the scientific reality that there are way too many people on this planet and that a plethora of issues actually stem from this overpopulation (such as overuse of natural resources, the social failings of megacities, mobility problems, etc.). During all my years of researching eco-justice issues, I have not once encountered a single eco-justice text or website engaging with this issue which is clearly central to our concerns. Others such as Bello (2013: 173–180) have written about why population growth is indeed such a massive problem. Gregory Bateson already stated quite a while ago that "the population explosion is the single most important problem facing the world today". Why? "The very first requirement for ecological stability is a balance between the rates of birth and death." (Bateson 2000: 500) According to calculations from many different scenarios planet Earth cannot sustain more than 3 billion people (high estimate) in the long run (see discussion in Latouche 2011: 150–157). Obviously, one of the reasons for not engaging with this issue is that it is very difficult, as Jean-Paul Besset has put it, to "not be progressive (as in 'uncritical progress believer') any more without becoming reactionary" (2005). The discourse of 'too many (foreign) people' is one that is emanating from the political far right and there is no way I want to have anything to do with this xenophobic approach. The problem is clearly not immigration or people less privileged than we are, forced to leave their home as environmental, political or war victims. Yet, there *are* way too many people on the planet as a whole, and there is overwhelming evidence that Paul Ehrlich is right with his formula: overpopulation is not good for the planet, and this means also: not good for our long-term survival.

I am bringing up all these issues because I am convinced that they ought to be of highest concern for anybody with an interest in eco-justice. As long as we still refuse to engage with these burning issues and refuse to let go of the dualistic, non-system-ic, non-scientific approaches sketched out above we will never even come close to an understanding of what is going on within and around us, let alone come up with solutions that actually deal with reality as opposed to some ideologically conceived notion of it.

Is modernity and science really the biggest obstacle to eco-justice?

Now you might say: all very well, I get it. But what does this have to do with eco-justice thinking and practice? Surely, we eco-justice practitioners are all on the same page here and are not in any way guilty of such non-systemic approaches. In a lot of cases, I would not want to argue with this. There are fantastic projects going on, such as ecojustice Canada, with their down-to-earth approach (http://www.eco-justice.ca/approach/, accessed 27 March 2016), or, similarly, the Eco-Justice Collaborative in Chicago (http://ecojusticecollaborative.org/, accessed 27 March 2016), to name but two examples. And even if we look closer to home, very close, such as the Eco-Justice Press website, there is, at first sight, nothing wrong. When we look at the five criteria which should guide submissions to the Press (http://ecojusticepress.com/ > Eco-Justice as a Guiding Conceptual and Moral Framework, accessed 28 February 2016) there is clearly nothing to be said against the vocal criticism of the abuse of the multinational chemical industry of natural resources, the criticism of western hyper-consumerism and the call for truly sustainable practices in all walks of life.

But as always: the devil is in the detail. We can see here and in other eco-justice texts and websites some sloppy thinking and ignorance at work. Underlying it all is a value-bias which makes me distinctly uneasy. As I shall argue below, scientific progress is nothing to be ashamed of, on the contrary. It is, together with the other achievements of modernity, namely democracy, equal rights, free speech, communities guided by law, etc., the only basis we have for an ever better and sounder understanding of life in all its dimensions and it is the only way to free us of the shackles I mentioned at the beginning. That science in the hands of powerful elites and corporations can lead to precisely the colonial destruction, exploitation and injustice we want to eliminate, is not an argument against scientific understanding but against power, greed and non-democratic structures—which links us back to what Chris Hedges said above about the fights we need to pick. But, once again, as the ISIS terrorist attacks in more than 20 countries (including in Paris, Baghdad, Beirut, Tripoli, Brussels, Tunis, as well as the Sinai Peninsula, Saudi Arabia, Yemen, Kuwait, Turkey, Russia, Afghanistan and Indonesia) have so clearly shown: only free, open and democratic modern societies allow us the safe space to argue about truth and values; the ideologies of the ISIS and other non-secular states and societies only know the language of violence and oppression.

I am very much concerned by the underlying damnation of the West and the correspondent idealization and romanticizing of 'indigenous' communities and traditional ways of knowing—as expressed in Eco-justice Press criteria 3 and 4 ("Revitalizing the cultural commons", "ecological traditions of earth democracy"). This strand—vilification of modernity, 'Western colonization' and the corresponding romanticizing of traditional and indigenous societies—is very strong in all the eco-justice material I encountered over the years, and it is deeply steeped in the post-modern fog that I tried to penetrate with Sokal above. *EcoJustice Education*, arguably one of the most important textbooks in the field, now in its second edition, displays this already in the introduction, where we read, for example: "While not "wealthy" by Western standards of material or political status, they were able to feed and shelter their families without external interference." (Martusewicz et al. 2015: 6). This implies that all would be well if modernity wouldn't have happened and interfered. But it goes on: the "Conceptual toolbox" for *EcoJustice Education* (http://cw.routledge.com/textbooks/9780415872515/toolbox.asp, accessed 28 March 2016) offers the following definitions, again clearly negatively connoting modernity and romanticizing indigenous communities and oral traditions of 'knowledge':

> "*Discourses of Modernity*: The specific set of discourses that together create our modern, taken-for-granted *value-hierarchized worldview*, including anthropocentrism, *progress*, individualism, *science/rationalism*, mechanism and so on."
> "*EcoJustice Education*: (…) is challenging the *deep cultural assumptions underlying modern thinking* (...); and the recognition of the need to *restore* the cultural and environmental *commons*."

versus, clearly positively meant:

> "*Indigenous knowledge*: *knowledge* that has been *passed down through generations* regarding how to *live successfully* in a particular place. It is generally *spiritually based* and includes a variety of interrelated dimensions: physical, biological, linguistic, spiritual, social, and economic."

> "*Indigenous People*: those peoples who predate any other groups living in a particular region, and who define themselves through a "*spiritual link to the land*".
> "*Oral traditions:* the *Indigenous* practice of passing on *knowledge* and *moral instruction* through verbal modes such as storytelling." (italics added)

There is not a hint of critical reflection here; that on the one hand modernity might also have its advantages, and on the other hand that there might be real issues with "knowledge passed down through generations", with oral "knowledge" and with the "moral instruction" mentioned, in terms of reliability, new insights over time, personal freedom and liberty, and, indeed, truth. This is by no means an exception: In an earlier special issue on "Ecojustice and Education" of the journal *Educational Studies* we find almost on every page statements like: the solution is the "revitalization of the commons (…) through the affirmation of wisdom that these people have had all along", there is a need to "preserve centuries-old knowledge" (Wayne & Gruenewald 2004: 3), the "desire to recognize and preserve traditional knowledge and cultural practices" (ibid.: 4), the "need to conserve cultural traditions" (ibid.: 56), the claim that "where [the West] has found traditional forms of knowledge, it has brought "Reason" and science" (ibid.: 59; note the quotation marks), or: "the vibrant soil beneath our feet becomes another victim of modern inattentiveness to Creation" (ibid.: 70; note the captial C), and that "the cultivation of these [ethical] qualities (…) will almost certainly require a relinquishment of modern assumption[s]" (ibid.: 91), and finally that being modern forces us "away from ancestral ways of being" and has led to "a devastating loss of the ecology of indigenous language" (ibid.: 124). But this uncritical acceptance of labels, such as "commons", "traditional knowledge" etc., can also be seen in the write-up for the "EcoJustice and Activism Conference", themed "Reclaiming the Commons: Diverse Ways of Being and Knowing", which took place in Michigan in March 2016:

> "We understand the commons as social and political, cultural and ecological, ontological and epistemological including *often-ancient practices*, relationships, *traditions*, *knowledge*, skills, and ways of being—both human and the *more-than-human*." (http:// ecojusticeconference.weebly.com/, accessed 27 March 2016; italics added)

The pre-conference program, incidentally, is full of meditation retreats, offering "meditation tools to support our ecojustice efforts (...) including: breathing meditation, mindfulness, mantra practice and contemplative movement meditation" or "Meditation on the *Four Immeasurables*" (http://ecojusticeconference.weebly.com/ program.html; italics added); in the program itself we find sessions on "Re-Membering Our *Lack* of *Sacred Cultural Ceremonies* in Modern Times" (poster session) or "The Knowledge of My Mother is My Knowledge" (https://drive.google.com/ file/d/0B8CDnY9sPS7BVGlURll0X2tVTVk/view?usp=sharing, accessed 26 March 2016).

Why traditional, indigenous or subjective everyday 'knowledge' won't help much

Why would all this be a problem? Despite the allure of simplicity and the yearning for innocent paradise, indigenous peoples or traditional knowledge can in most cases help us only in very limited ways to solve our current problems—if we really try to assess its validity for current issues and the truth value beyond generalizations. It means, however, that we are not fooled by superficial slogans such as "we are all children of Mother Earth" or "valuing the Elders". Fact is that their cosmologies, their attempts to understand the world were not somehow better, more intuitive or more holistic than ours. In retrospect we know that it simply was a very limited way of understanding the complex world around them, severely restricted by the fact that they did not know much about how the world really works and came about. If you probe deeply and critically into their ways of understanding, you start to realize these limitations. Take a look at the way they understood human bodies and health. Talking about meridians, for example, or about chakras etc. is owed to the fact that within these cultures nobody dared to do what Leonardo da Vinci finally dared, namely cut open corpses to see what was really going on inside a body. As long as you don't do this, you have nothing left but your imagination of what might be. You have no way of checking, of distancing knowledge from the immediate assumptions and intuition at first sight. This leads precisely to the metaphysical explanatory systems, i.e. social constructions of 'meaning', which invent and narrate 'understanding' in the absence of procedures and experimentally tested and testable theories to verify those tales. The people then did not have any tools and scientific instruments at hand to peak beyond their everyday experience and traditional ways of making sense of this dangerous, unpredictable and frightening inside and outside world. I don't in any way want to belittle them for this lack of understanding: they might well have had the best possible explanatory systems at those historic moments in time. But today, if we were to do our homework and really dig into the accumulated knowledge of today's world, we should know better. And there is another point we should not forget: the 'explanatory' tales and sacred, traditional 'knowledge' had often a lot more to do with confirming power structures and social standing of the ones 'in the know' (medicine man, wise woman, elders, bearers of the secret/sacred truth etc.) within those societies than the endeavor to seek the truth.

In other words, rather than being scientifically sound these types of 'knowledge' rely either on dogma (the particular tribe 'knows' since the beginning of time that such-and-such is the truth/wisdom, and you are not allowed to question it, for fear of being expelled from the community—which is why these belief systems often work like sects), or on everyday understanding. But, as Werner Obrecht has made clear, everyday knowledge, our immediate and unreflected experiences, are very rarely suitable for generating sound knowledge:

> "Everyday thinking does not understand itself, is therefore uncritical and, if at all, only partially able to come to true statements. (…) Without critical theory of itself and without a theory of the nervous system everyday thinking (…) believes that it understands the material things in the world directly as they are (naïve realism). (…) Its implicit metatheory is equal to the one in magical and religious worldviews and is source of resistance against the scientific world-view of adults." (Obrecht 2009: 56)

Or, provocatively put: "Every political or ethical argument which is drawn from a specific experience is wrong." (Gilles Deleuze, quoted in Žižek 2016)

Why is this so? Our personal experience and 'knowledge' is so shaped by culture, unreflected belief systems (such as those handed down to us by our parents), myths, taken-for-granted assumptions embedded in our language, the structure of our social and economic reality etc., that assuming it could easily generate something resembling a reliable truth is more than a little naïve. And traditional societies and indigenous communities had not much more at their disposal than a sort of amalgam of everyday thinking and experiences of their members. They did not have at their disposal what I call the scientific mindset which is crucial and non-negotiable if we are to understand the world we live in. A good indicator that this is the case can be provided by a look into history: more and more of the traditional belief systems and explanations prove to be wrong, the more we advance with scientific understanding (think of 'flat earth', 'earth = center of the universe', 'creator'). This is because they were essentially claims and guesses which slowly but surely are replaced by sound understandings. In other words: they were not explanations but often elaborately narrated markers of our ignorance, i.e. the fact that we simply did not know yet.

Another important point in this context: if you really analyze the ways these in our circles so often romanticized traditional or indigenous communities work you will quickly find that their understanding of human beings, of power, of gender is such that nobody today would freely decide to live in such autocratic, dogmatic, often cruel, sexist and racist communities. As a philosopher and historian I must say that it is no surprise that none of these societies ever produced anything resembling an open, democratic society worthy of that name. This is no wonder: they rest on very rigid belief systems and moral codes which are not open to adaptation or even rejection or dismantling, a fact, of course, which is also true for religious communities (see below).

To my mind, we really have to own up to the fact that there is no alternative to the open, transparent, (self-) critical approach of scientific understanding which is the exact opposite of 'belief' of any kind (which denies criticism and doubt, thereby making true understanding and learning impossible). There is not a single field of scientific knowledge today which can be encapsulated in a single book which then 'stays true'. Yet traditional, indigenous and religious/spiritual belief systems (so positively evoked in the eco-justice quotations above) often rely on a single text or foundational myth—in indigenous contexts often claimed to be 32'000-year-old wisdom. With just a little knowledge about evolution, history and the change of individuals over their lifetime and societies over time, or paradigm changes in knowledge systems, we know that it is simply impossible that they can hold much truth about us today. Scientific knowledge is precisely replacing with new understandings what we long thought to be 'true'.

The self-critical openness to being proven wrong (i.e. learning): the scientific approach

But what do I mean exactly by a scientific approach? Let me quote Popper:

> "But science is one of the very few human activities—perhaps the only one—in which errors are systematically criticized and fairly often, in time, corrected. This is why we can say that, in science, we often learn from our mistakes, and why we can speak clearly and sensibly about making progress there." (Popper 1963)

> "When I speak of reason or rationalism, all I mean is the conviction that we can learn through criticism of our mistakes and errors, especially through criticism by others, and eventually also through self-criticism." (Popper 2001)

> "The game of science is, in principle, without end. He who decides one day that scientific statements do not call for any further test, and that they can be regarded as finally verified, retires from the game." (Popper 2004)

When I talk about the scientific mindset I talk about this open approach which will never accept any dogma, any claim by the powerful, the elites, tradition or anybody else at face value. It always wants proof, explanation, evidence, and always independent of the person that tries to impose a certain truth. Only if insights are replicable independently, if they work outside a specific community is there a chance that we are onto something which is beyond the illusions, the traditional 'this is what we always believed' and the self-justifications of individuals, communities and groups with vested interests: "It's this demonstrability and repeatability that makes science unique: it requires no indoctrination to accept." (Spadafino 2016) This approach is the only one which allows us to liberate ourselves from the crutches of fear, of ignorance, of 'eternal truth' being forced upon us. Or to quote Kant again: we need to emerge "from [our] self-imposed immaturity" (Kant 1784). True to its enlightenment origins, we should not underestimate or willfully neglect the liberating power of science:

> "Science flings open the narrow window through which we are accustomed to viewing the spectrum of possibilities. We are liberated by calculation and reason to visit regions of possibility that had once seemed out of bounds or inhabited by dragons." (Dawkins 2007: 418)

Of course, there is an arrogance and destructive side to knowledge as well, when science forgets Popper's humble insight that "the game of science is, in principle, without end", when it abuses knowledge for the gain of economic and political power. But still, this game is the only game we can play to guarantee not just freedom, but also justice, understanding and the protection from those who want to force us to believe things without any attempt, let alone the ability, to prove what they claim. So the science game, however prone to abuse and imperfections it may be, is the only game in town.

Have we, truly, lived up to these insights in the eco-justice movement? Have we truly focused the same sharp criticism that we levelled against modernity, progress and rationalism onto our own favorites such as 'inter-generationally renewed traditions' or 'spirituality'? Unless we start to shed all sorts of myths, half-truths and historically, sociologically and economically unjustifiable illusions and idealizations about 'the commons' or about traditional and indigenous 'knowledge' we will be caught in an inability to understand what is really going on and what is needed. Or—with a view to climate change or creationism—it doesn't matter in any way what we belief or which opinion we hold. It only matters what today can collectively, and independently, be established as truth: "science is not there for you to cherry pick" (Neil deGrasse Tyson in Weathers 2014).

… and then there is religion…

But wait, we are not done yet. There is an even bigger problem we have in the eco-justice movement, connected to the biggest taboo we have worldwide in the face of moving towards a proper understanding of the three dimensions of power indicated by Horkheimer and Adorno, and towards truth in general: religion.

More than half the websites which come up if you search for eco-justice marry it with religion[1]. Often, these sites use an image which marries ecology, justice and faith (see http://ecojusticenow.org/page20/page20.html). And you can read sentences like these, which make you wonder in which century we are living: "Science, by its very nature, cannot offer enough guidance for the challenges of contemporary environmental policy." (Holmes Rolston III, http://ecojusticenow.org/resources/Eco-Justice-Ethics/HLPR-Saving-Creation.pdf, accessed 28 March 2016). Even more worryingly, the term eco-justice itself seems to come from a religious context: "To foster converging commitments to ecology and justice, American Baptist leaders Richard Jones and Owen Owens introduced the term eco-justice." (Hessel 2007) Hessel claims that from these beginnings "within two decades, a significant body of writings emerged that emphasize respect for every kind and show intersecting concern for ecology, justice and faith" (ibid.). His "Environmental Justice Annotated Bibliography" offers a key insight into how strongly the development of eco-justice has been influenced by religious environmentalists. This seriously means that we are in dire straits. It truly is, even in the days of Baghdad, Paris, Tripoli and Brussels, still a total taboo to point the finger at possibly *the* most influential source of unreason and *the* prime example of a blatant disregard for truth worldwide, far more influential and damaging than the intellectual fog that postmodernism, multiculturalism and relativism pulled over our eyes. Unless we acknowledge this and take issue with all sorts of religious beliefs and belief-residues also in the environmental and eco-justice movement, we have simply

---

1 _____ For example Eco-Justice Ministries (http://www.eco-justice.org/), The Forum on Religion and Ecology at Yale (http://fore.yale.edu/disciplines/ethics/eco-justice/), http://www.episcopalchurch.org/page/eco-justice, Racine Dominican Eco-justice Center: "Commitment to truth in the light of the Gospel" (http://www.racinedominicans.org/eco-j.cfm), eco-justice camp, sponsored by the Unitarian Universalist Church of Palo Alto (UUCPA) (http://ecojusticecamp.com), EcoJusticeNow (http://ecojusticenow.org/page20/page20.html), World Council of Churches (https://www.oikoumene.org/en/what-we-do/eco-justice), National Council of Churches of Christ Eco-Justice Programs "justice for God's planet and God's people" (https://ecojustice.wordpress.com/), Interreligious Eco-Justice Network (http://irejn.org/), all accessed 15 March 2016).

no chance in hell to ever get to an open, humane, democratic and just society which respects the limits of the biosphere.

This is the kind of toughness in arguments which we need to relearn, according to Žižek, even if—as likely in this case—we are hurting people's feelings. But: "Their concern, their pain is no measure for the truth." (Žižek 2016) Speaking as a historian again, it is difficult to name a force in history which has been and still is more destructive than religion (with the possible exception of its bed-fellow capitalism); it is impossible to name a historical force that is more responsible for the spread of ignorance than religion (just look at Catholicism and its dogmas on sexual relationships, contraception and abortion, for instance); it is impossible to name a force in history which has done more harm to women and their standing in society than religion; the list could go on and on. And on top of all that, there is good scientific evidence that religious people, when it comes down to how they behave rather than how they *think* they behave, are more opinionated, less altruistic, less peaceful, more prone to violence, to exorcising others from their group etc.; in other words that they have a very screwed-up morality which would wreak havoc to any modern sense of justice and morality (see Dawkins 2007: 258-348; Dawkins' book is in any case a wonderful wholesale debunking of every myths about religion you ever believed in).

And, of course, religious world-views, their understanding of the human being, of evolution, of how the universe works have nothing whatever to do with what we know today about these issues—quite apart from the fact that religion, historically, has always been what Marx claimed, namely "opium for the masses", i.e. a wonderful tool of oppression for the powerful: "Religion is excellent stuff for keeping common people quiet" (Napoleon, quoted in Dawkins 2007: 313). So there is little more to say than what Albert Einstein and Salman Rushdie have stated:

> "The word God is for me nothing more than the expression and product of human weakness, the Bible a collection of honorable, but still purely primitive, legends which are nevertheless pretty childish. No interpretation, no matter how subtle, can change this for me. For me the Jewish religion like all other religions is an incarnation of the most childish superstition." (Einstein 1954)

> "Religion [is] a medieval form of unreason." (Salman Rushdie in: The Guardian 2015)

Superstition and unreason: This is what Latouche means by "auto-immunization" of religion (2011: 192): you can only believe in religious dogma by "building up irrational attitudes of submission to authority" (Chomsky 1992). Religious belief can never withstand the test of independent, external, objective or inter-subjective scrutiny. And more and more of these religious "yes, but how do you explain ..." fall by the wayside with the advances of our understanding of the universe: Christopher Hitchens, in *The Portable Atheist*, collects a couple of essays from leading scientists who show that modern Physics has no need any more for an external creator to explain the coming into being and existence of our world (see Victor Stenger [311–327], Steven Weinberg [366–379] in Hitchens 2007). And Capra/Luisi have beautifully shown that we don't need any 'external help' to explain life on Earth: all you need is an understanding of the processes of emergence and self-organization (Capra/Luisi 2014: 180-181, 226-261). As Castoriadis says in the motto of this essay and Kant urges us in his challenge to "emerge from [our] self-imposed immaturity", it takes a bit of courage to understand and acknowledge that there is no plan, no aim, no teleology to nature and evolution; life is the product of pure chance. Once this is understood the fear which drives every religion and superstition just falls away. Yet, we do not even seem to try to acquire this understanding. Dawkins rightly says that "one of the truly bad effects of religion is that it teaches us that it is a virtue to be satisfied with not understanding" (quoted in Hitchens 2007: 297).

Maybe Salman Rushdie put it most clearly, in his beautiful "Letter to the Six Billionth World Citizen", entitled "Imagine There's No Heaven":

> "To choose unbelief is to choose mind over dogma, to trust in our humanity instead of all these dangerous divinities. So, how did we get here? Don't look for the answer in 'sacred' storybooks. (...) The ancient wisdoms are modern nonsenses. Live in your own time, use what we know, and as you grow up, perhaps the human race will finally grow up with you and put aside childish things. (...) Once and for all, we could put the stories back into the books, put the books back on the shelves, and see the world undogmatized and plain. Imagine there's no heaven, my dear Six Billionth, and at once the sky's the limit." (quoted in Hitchens 2007: 382–383)

Conclusion: do we walk the talk? Do we firefight where the fire is?

All this should give us plenty of reason for self-criticism: Have we, in the eco-justice movements around the world, self-critically evaluated our beliefs and mental models to check if they are demystified and open to reasoning, sound knowledge and self-criticism? Have we really, truly moved beyond religion, beyond superstition and unreason, finally arriving in the reality of a secular 21st century? (But then again, this is a difficult one: since ideology and religion truly are forms of unreason, believers are rarely ever open to reasoned arguments. Therefore, they usually try to shoot the messengers so that they don't have to deal with the message.)

But even if you cannot follow me on this debunking of religion, let's just look at our own actions:

> "The experience that we have of our lives from within, the story we tell ourselves about ourselves in order to account for what we are doing, is fundamentally a lie—the truth lies outside, in what we do." (Žižek 2008: 47)

Another important aspect of religion, ideology and non-scientific explanatory systems is that they focus very much on what one believes, or—in Žižek's words—the stories we tell ourselves about who we are, what is right and wrong etc. The point today, in times of climate change and non-sustainability, is that this is almost irrelevant. These narratives are, by and large, excuses and lies, since it only matters what real-world impact we have on the biosphere. Only our actions count, however we might explain or justify them. As Niko Paech, one of the foremost post-growth economists in Germany, put it: "The time for excuses is over" (Paech 2014), meaning: we have more than enough knowledge, tried and tested social models, technologies and tools at our disposal to live sustainably. The only reason why we don't do it, is because we always find stories and narratives and excuses to *not* live the eco-justice option. Hand on heart: how many of us who claim to be truly convinced eco-justice fighters, are actually living an eco-just life? How many of us have ditched flying? How many have sold our car? How many live exclusively off renewable energy? How many of us are only feeding ourselves from locally grown, organic food? And how many of us are really fighting Chris Hedges' fights?

There is all the scientific evidence needed out there to show us that we can only move towards an eco-just, non-exploitative future by reducing our impact on planet earth dramatically. This unfortunately means: we have to actually *do* it! It doesn't matter how we justify and explain it: if our daily actions are not sustainable, even if we *think* they are, it doesn't help.

To summarize: given the immense and complex problems of power, wealth, exploitation and oppression, we need the best, most accurate knowledge there is to solve them, from a wide range of disciplines. What we most often tend to do in our circles, though, is to give in to the craving to wish us back into a pre-modern simplicity where wishful thinking was thought to help. No: let us firmly arrive in the 21st century. Or, in other words: if we want eco-justice to succeed we need five things:

- we need to remember Horkheimer and Adorno and fight oppression on all levels: within ourselves, between us and as exploitation of nature;

- we need to take Kant seriously and truly have the courage to use our own mind "without guidance from others";

- we need to be absolutely sure that our arguments are scientifically sound and watertight, and don't rest on superstition, unreason or belief;

- we need to be the role models in everything we *do*; the preaching to others we can leave to our brothers stuck in medieval mindsets;

- we need to be sure that we pick the right friends (science rather than unreason) and the right enemies (corporations, power, wealth rather than science, modernity and the enlightenment).

❖ ❖ ❖

Dr. Rolf Jucker is currently Director of SILVIVA, the Swiss Foundation for Experiential Environmental Education (http://www.silviva.ch). From 2013 till 2014 he was a teamleader at Kaospilots Switzerland (www.kaospilots.ch). Until the end of 2013, he ran the Department of Conceptual ESD Development at the Swiss Foundation for ESD (www.education21.ch). From 2008 to 2012, he was Director of the Swiss Foundation for Environmental Education (www.umweltbildung.ch), an independent body tasked with the mainstreaming of environmental education and ESD in the Swiss school system. Having gained an MSc in Environmental and Global Education at London South Bank University, he has worked extensively as an international ESD advisor and has widely published on environmental education and justice issues (see rolfjucker.net), amongst others *Our Common Illiteracy. Education as if the Earth and People Mattered* (2002) and *Do We Know What We Are Doing? Reflections on Learning, Knowledge, Economics, Community and Sustainability* (2014).

References

Bateson, Gregory (2000). *Steps to an Ecology of Mind.* Chicago: The University of Chicago Press. (Original 1972).

Beckert, Sven (2014). *Empire of Cotton. A Global History.* New York: Alfred A. Knopf.

Bello, Walden (2013). *Capitalism's Last Stand? Deglobalization in the Age of Austerity.* London; New York: Zed Books.

Besset, Jean-Paul (2005). *Comment ne plus être* progressiste … *sans devenir* réactionnaire. Paris: Librairie Arthème Fayard.

Brown, Brené (2013). *Daring Greatly. How the Courage to Be Vulnerable Transforms the Way We Live, Love, Parent and Lead.* London: Portfolio Penguin.

Capra, Fritjof, and Pier Luigi Luisi (2014). *The Systems View of Life. A Unifying Vision.* Cambridge: Cambridge University Press.

Castoriadis, Cornelius (2005). *Une société à la dérive. Entretiens et débats 1974-1997.* Edition préparée par Enrique Escobar, Myrto Gondicas et Pascal Vernay. Paris: Editions du Seuil [=éditions points 650].

Chomsky, Noam (1992). Excerpts from the film Manufacturing Consent. Interview. https://chomsky.info/1992____02/. Accessed 26 March 2016.

Chomsky, Noam, and Edward S. Herman (1994). *Manufacturing Consent. The Political Economy of the Mass Media.* London: Vintage.

Crockett, Christopher (2016). There's far more to the galaxy than meets the eye. In: *Science News*, Vol. 189, No. 8, April 16, 2016, p. 36. https://www.sciencenews.org/article/theres-far-more-galaxy-meets-eye?utm_source=Society+for+Science+Newsletters&utm_campaign=aa1264b189-SN_Editor_s_Picks_April_4_20164_8_2016&utm_medium=email&utm_term=0_a4c415a67f-aa1264b189-104656677. Accessed 23 April 2016.

Ehrlich, Paul (2013). I = PAT. http://en.wikipedia.org/w/index.php?oldid=502224816. Accessed 26 March 2016.

eia (U.S. Department of Energy) (2013). International Energy Statistics: Total Primary Energy Consumption. http://www.eia.gov/cfapps/ipdbproject/IEDIndex3.cfm?tid=44&pid=44&aid=2. Accessed 26 March 2016.

Einstein, Albert (1954). Letter to Erik Gutkind, 3.1.1954. http://www.lettersofnote.com/2009/10/word-god-is-product-of-human-weakness.html. Accessed 26 March 2016.

Graeber, David (2011). *Debt. The First 5000 Years*. Brooklyn, NY: Melville House.

Haraway, Donna J. (2008). *When Species Meet*. London, Minneapolis: University of Minnesota Press.

Hedges, Chris (2010). *The World as it is: Dispatches on the Myth of Human Progress*. New York: truthdig / Nation Books.

Hessel, Dieter T. (2007). Eco-justice Ethics. Environmental Justice Annotated Bibliography. Part of: *The Forum on Religion and Ecology at Yale*. http://fore.yale.edu/disciplines/ethics/eco-justice/. Accessed 25 March 2016.

Hitchens, Christopher (Ed.) (2007). *The Portable Atheist. Essential Reading for the Nonbeliever*. Philadelphia: Da Capo Press / Perseus Books.

Horkheimer, Max, und Theodor W. Adorno (1986). *Dialektik der Aufklärung. Philosophische Fragmente*. Mit einem Nachwort von Jürgen Habermas. Frankfurt/M.: S. Fischer (original 1944).

Huxley, Aldous (1989). *Brave New World*. New York: HarperPerennial.

Jucker, Rolf (2014). *Do we know what we are doing? Reflections on learning, knowledge, economics, community and sustainability*. Newcastle upon Tyne: Cambridge Scholars Publishing. http://www.cambridgescholars.com/do-we-know-what-we-are-doing-reflections-on-learning-knowledge-economics-community-and-sustainability

Kant, Immanuel (1784). Beantwortung der Frage: Was ist Aufklärung? In: *Berlinische Monatsschrift*, December 1784. https://de.wikisource.org/wiki/Beantwortung_der_Frage:_Was_ist_Aufkl%C3%A4rung%3F. Accessed 25 March 2016.

Keith, Lierre (2009). *The Vegetarian Myth. Food, justice, and sustainability*. Crescent City, CA: Flashpoint Press; Oakland, CA: PM Press.

Latouche, Serge (2011). *Vers une société d'abondance frugale. Contresens et controverses sur la décroissance*. Paris: Mille et une nuits [Les Petits Libres No. 76].

Leiva, Steven Paul (2012). The Arrogance of Ignorance—The Authority of Knowledge. In: *The Huffington Post Arts & Culture*, 02.10.2012. http://www.huffingtonpost.com/steven-paul-leiva/the-arrogance-of-ignoranc_b_1930595.html. Accessed 26 March 2016.

Lindquist, Kristen A. (2016). What can the brain tell us about emotion? In: *Emotion Researcher. ISRE's Sourcebook for Research on Emotion and Affect*. http://emotionresearcher.com/the-emotional-brain/lindquist/. Accessed 24 March 2016.

LPR (2012). Living Planet Report 2012: Biodiversity, biocapacity and better choices. http://www.footprintnetwork.org/images/uploads/LPR_2012.pdf. Accessed 26 March 2016.

Lummis, C. Douglas (1996). *Radical Democracy*. Ithaca; London: Cornell University Press.

Martusewicz, Rebecca A., Jeff Edmundson and John Lupinacci (2015). *EcoJustice Education: Toward Diverse, Democratic, and Sustainable Communities*. 2nd Edition. New York: Routledge.

New Scientist (2013). How long will it last? http://www.newscientist.com/data/images/archive/2605/26051202.jpg. Accessed 26 March 2016.

NYT (New York Times) (2013). Heat-Trapping Gas Passes Milestone, Raising Fears. http://www.nytimes.com/2013/05/11/science/earth/carbon-dioxide-level-passes-long-feared-milestone.html?pagewanted=all&_r=0. Accessed 26 March 2016.

Obrecht, Werner (2009). Die Struktur professionellen Wissens. Ein integrativer Beitrag zur Theorie der Professionalisierung. In: Roland Becker-Lenz *et al.* (Eds.): *Professionalität in der Sozialen Arbeit. Standpunkte, Kontroversen, Perspektiven* (pp. 47-72). Wiesbaden: VS Verlag für Sozialwissenschaften.

Orwell, George (1949). *1984*. New York: Signet Classic.

Paech, Niko (2014). Konsum nervt. In: *taz—die tageszeitung*, 1.9.2014. https://www.taz.de/!5034306/. Accessed 26 March 2016.

Popper, Karl (1963). *Conjectures and Refutations: The Growth of Scientific Knowledge*. London: Routledge. Source: http://en.wikiquote.org/wiki/Karl_Popper. Accessed 26 March 2016.

Popper, Karl (2001). On Freedom. In: Karl Popper: *All Life is Problem Solving*. London: Routledge. Source: http://en.wikiquote.org/wiki/Karl_Popper. Accessed 26 March 2016.

Popper, Karl (2004). *The Logic of Scientific Discovery*, Chapter II, Section XI. London: Taylor & Frances [Routledge Classics]. (Original 1934). Source: http://en.wikiquote.org/wiki/Karl_Popper. Accessed 26 March 2016.

Rockström, Johan *et al.* (2009). Planetary Boundaries: Exploring the Safe Operating Space for Humanity. In: *Ecology and Society*, 14(2), 32. http://www.ecologyandsociety.org/vol14/iss2/art32/ . Accessed 26 March 2016.

Rose, Steven (2006). *The 21st-Century Brain. Explaining, mending and manipulating the mind*. London: Vintage Books.

ScienceDaily (2012). Record High for Global Carbon Emissions. http://www.sciencedaily.com/releases/2012/12/121202164059.htm. Accessed 26 March 2016.

Sokal, Alan (1996). A Physicist Experiments With Cultural Studies. In: *Lingua Franca*, May/June 1996, pp. 62-64. http://www.physics.nyu.edu/sokal/lingua_franca_v4/lingua_franca_v4.html. Accessed 28 March 2016.

Sokal, Alan, and Jean Bricmont (1998). *Intellectual Impostures. Postmodern philosophers' abuse of science*. London: Profile Books.

Spadafino, Joseph T. (2016). Americans' Unwillingness to Accept Evolution En Masse Is a Failure of Science Education. *Huffpost Education: The Blog*. http://www.huffingtonpost.com/joseph-t-spadafino/americans-unwillingness-t_b_9693130.html. Accessed 16 April 2016.

The Guardian (2015). Salman Rushdie ready to 'get medieval' with the Times Literary Supplement. Novelist hits back at TLS blogger's criticism of his language in response to Charlie Hebdo attacks. In: *The Guardian*, 26.1.2015. http://www.theguardian.com/books/booksblog/2015/jan/26/salman-rushdie-get-medieval-times-literary-supplement-tls. Accessed 26 March 2016.

Wayne, Kathryn Ross, and David A. Gruenewald (2004). *Special Issue: Ecojustice and Education*. In: *Educational Studies. A Journal of the American Educational Studies Association*. Vol. 36, No. 1, August 2004.

Wardsauto (2015). World Sales Hit Record High in March. 30.4.2015. http://wardsauto.com/industry/world-sales-hit-record-high-march. Accessed 26 March 2016.

WCA (2014). World Coal Association: Coal Statistics. http://www.worldcoal.org/sites/default/files/coal_steel_facts_2014(12_09_2014).pdf. Accessed 26 March 2016.

Weathers, Cliff (2014): Neil deGrasse Tyson Chastises Media For Giving 'Flat Earthers' Equal Time in the Climate Change Debate. The host of Cosmos: A Spacetime Odyssey says 'science is not there for you to cherry pick. In: *AlterNet*, 10 March 2014. http://www.alternet.org/environment/neil-degrasse-tyson-chastises-media-giving-flat-earthers-equal-time. Accessed 26 March 2016.

Wilkinson, Richard & Pickett, Kate (2010). *The Spirit Level. Why equality is better for everyone*. London: Penguin Books.

Worldwatch (2013a). The State of Consumption Today. http://www.worldwatch.org/node/810. Accessed 26 March 2016.

Worldwatch (2013b). Global Meat Production and Consumption Continue to Rise. http://www.worldwatch.org/global-meat-production-and-consumption-continue-rise-1. Accessed 26 March 2016.

Zinn, Howard (1996). *A People's History of the United States. From 1492 to the Present*. 2nd edition. London, New York: Longman.

Žižek, Slavoj (2008). *Violence*. New York: Picador.

Žižek, Slavoj (2016). "Liberal? God help!" Slavoj Žižek on new walls, prosperity and world revolution. An interview with René Scheu. In: *Neue Zürcher Zeitung, Literatur und Kunst*, Saturday, 30 January 2016, pp. 54-55.

RE-IMAGINING EDUCATION FOR ECO-JUSTICE:
THROUGH THE LENS OF SYSTEMS THINKING, COLLECTIVE INTELLIGENCE AND CROSS-CULTURAL WISDOM

Thomas Nelson & John A. Cassell

Climate change and environmental degradation are the most pressing issues facing humanity in the 21st century. We are living beyond the planetary carrying capacity and efforts to seriously address sustainable practices remain remote at best, and clearly lacking in political leadership. "If indeed the very survival of human civilization is directly linked to the survival of the natural world and its myriad species, then it is imperative that educators assume some sense of responsibility" (Nelson & Coleman, 2012, pg. 155.).

Coming to grips with the state of eco-justice in the geopolitical, economic and globalized environment of today is a rather daunting affair. This is made even more complex by the fact that a thick patina of now arcane neoliberal thinking rooted in a nineteenth century mindset has deeply penetrated the intellectual and operative dynamics of the post-industrial West. To contemplate the enormity of the cumulative effect of human activities over the span of our species' time on Earth with regard to the intentional transformation of the chemical composition of the biosphere and everything on it is truly an overwhelming task. Without question the future of the inherent diversity of all living things, including the human species, is imperiled by a rapidly changing climate directly linked to the activities and by-products of a civilization that has failed to recognize planetary resource limitations. Hegemonic intellectual formulations that drive the social and economic ordering of Western civilization are hopelessly out of alignment with the current facts on the ground, and the simple fact that allowing aggression, war, and the forceful taking from others that which is deemed immediately important and economically valuable to remain the primary process by which problems and differences are solved is no longer sustainable in a world of increasing and potentially catastrophic scarcity. The increasingly unavoidable reality is that this ultimately vicious expression of social Darwinism is a dark luxury that the world community can no longer afford.

The Nature and Scope of the Problem

As the continually growing production cycle of fossil fuels and synthetic chemicals interact and combine into myriad forms and are then absorbed into the planetary fabric, the ability for living organisms to adapt to an ever-more unnatural transformation of the system is increasingly at risk. The genetic map of life itself is being re-configured in ways that portend an even darker future ahead. McKibben (2011) compels us to consider the reality that we are now living on a completely different planet than the one that existed prior to the industrial age, born from the fuel of ancient fossils. The compelling question that follows is one of whether the foundational a priori assumptions upon which Western civilization rests are aligned with that all-compelling reality that McKibben points out and, further, one must consider whether the Western mindset is capable of successfully undertaking that alignment without meaningful contributions of other socio-cultural insights and the "lifeways" which emerge from them (McLaren & Ryoo, 2012).

Perhaps the ultimate irony in all this is that we are living in an age in which we find ourselves confronted by the single greatest threat yet encountered in our long sojourn on the planet --- a threat of our own making. Mounting scientific evidence confirms that the effects of the forms of human-environmental interaction that have accompanied the global expansion of the Western industrialized and post-industrial consumer driven economy are, in fact, contributing to the degradation of Earth's ecosystems upon which we depend for our continued existence (Nelson & Cassell, 2012).

Humans are having an enormous effect on the health of both social and natural habitats worldwide. This seems to portend a grim future in light of the lack of any evidence suggesting that life on Earth is indeed improving in overall health and sustainability, or even that political unrest and military conflict is in any way easing. The human population has risen from approximately 1.5 billion at the turn of the 20th century to what is expected to climb to over 9 billion around the year 2050. The very nature and historical pattern of human demand for natural resources suggests a completely unsustainable future. When taken in conjunction with the continuing power and motive force of neoliberal political and economic structures that serve to compel and feed a hyper capitalistic and consumer oriented global culture, this state of affairs offers little hope for stabilizing civilization from its increasingly fractured and divisive orientation to the future and to the future of generations to come.

The rapidly growing need for more and more resources to sustain human civilization is resulting in an Earth that is rapidly warming and increasingly compromised by the abundant introduction of dangerous synthetic chemical compounds and emissions from fossil fuel burning. This is quite literally altering the chemical composition of the planet. Polar ice is melting and the Arctic Ocean, once covered in sea ice reflecting sunlight back into the atmosphere, is thinning to the point of disappearing all together in coming decades. Glaciers are retreating at a pace that is almost mind-numbing. Ocean levels are rising faster than previously predicted. Island nations are losing critical land mass and some coral atolls are sinking all together. Ocean acidification has been inexorably exacerbated by increasing levels of carbon dioxide absorption, resulting in global coral bleaching and coral reef death. The Great Barrier Reef is being significantly compromised and may soon collapse completely, and with it innumerable marine species within one of the largest intact ecosystems on Earth. All the while tropical rainforests, the lungs of the planet, are being leveled to make room for food crops that will travel thousands of miles to dinner tables across the global landscape.

This state of affairs is requiring large populations of people, animals, and plants to relocate or migrate to higher ground, to latitudes unfamiliar. Those that cannot migrate and adapt to more hospitable climes will go extinct. Some say between 25 and 125 species of life every day are disappearing forever. The potentially disastrous fact of the matter is that biological diversity, a critical indicator of the health of any ecosystem, is being significantly compromised by a host of increasingly deleterious human activities, such as industrial monoculture agricultural practices and the resultant increasing volumes of toxic, cancer causing pesticides and fertilizers; mass deforestation; hydraulic fracturing; and the drawing down of ocean fish stocks due to fishing practices that wreak havoc on natural ecosystems. The earth is being stressed in ways never before seen. Natural habitats are rapidly deteriorating and in many places becoming void of life all together. Anthropogenic Climate Disruption (ACD) is contributing to what some refer to as the sixth mass extinction.

As alluded to above, toxins introduced into the environment through modern industrial development are causing cancer rates to rise significantly and adversely affect the rate and nature of human birth defects, mostly among the poor. The heaviest industrial chemical production and waste streams occur in poor communities, further reducing chances for social and economic advancement.

There is overwhelming agreement among the world's scientific community that human activities are directly associated with a rapidly warming planet and that we are on the brink of a point of no return where irreparable damage is imminent without a coordinated global effort to drastically reduce greenhouse gas emissions. The evidence of this building consensus is clear.

According to recent reports from NASA and NOAA, 2016 is poised to reach the warmest global temperatures in recorded history, far outpacing recent years, and spiking from the previous record global temperatures recorded in 2015. Most of the additional warming is due to accumulating greenhouse gases, including ever-increasing rates of methane emissions in the northernmost latitudes due to thawing tundra and the warming of Artic ocean water. In light of the continuously accumulating evidence from around the world, the scientific community overwhelmingly agrees that the Earth is already in the midst of an ecological crisis of staggering proportions, and will suffer extreme and irreparable damage without immediate efforts on a global scale to reduce greenhouse gas emissions.

This potentially catastrophic situation requires a re-examination of the deeply embedded and increasingly arcane foundational a priori assumptions upon which this increasingly disastrous course of events is predicated. This, in turn, requires that Western civilization retrace the intellectual steps that have brought us to this perilous precipice. We must deeply examine the hegemonic formulations upon which the neoliberal vision of life on Earth rests and consider the ways in which it has served to disfigure the human-environmental relationship. Part of this challenge is examining the role that institutionalized education has played in generating these phenomena and the role it might play in successfully addressing the lamentable results they have wrought. In doing so, educators must work to destabilize deleterious orthodoxies of the neoliberal mindset that have worked to degrade and endanger the natural environment. Taking on this challenge will necessitate the serious examination of the nature and meaning of social justice in the twenty-first century, how this concept relates to that of eco-justice and, further, what institutionalized education can do to create habits of mind that support and further both in ways that create a basis for confronting and addressing the neoliberal engine of destruction that has left numerous environmental crises at our doorstep.

The fact of the matter is that the Earth is on the brink of ecological collapse, primarily due to human activities that only continue to increase in scale and magnitude and portend a dire future for our children and their children's children. Of even more importance is the fact that these activities result from deeply rooted value systems and the hegemonic intellectual formulations and a priori assumptions that underlie them—a deeply embedded and densely interwoven subliminal cognitive web which is driving the post-industrial West—and the world it now politically and economically dominates—over the proverbial cliff. Tragically, confronting, assessing and reconsidering these foundational motive forces as part of seriously reflecting upon the lives and the kind of planet ecosystem we are leaving for future generations when making deeply impactful policy decisions does not appear to be a trait valued by the short sighted political and economic considerations at the center of the neoliberal agenda. The authors advance the position here that a way around this implacable intellectual roadblock must be found if humanity is to survive on planet Earth.

Neoliberalism and Institutionalized Education

The economic reward for the few in dominating the natural world and re-arranging the chemistry of planetary systems appears to be a mission largely driven by disturbing notions and claims of perpetuating the "advancement" of human civilization. This normative ideology is rooted in the belief that creating and implementing various technologies along a trajectory of ever-increasing complexity and ever-more comprehensive control over the natural environment aimed, ultimately, at an ever-expanding degree of resource extraction in support of an endlessly more lavish lifestyle for some at the expense of others is directly linked to advancing civilization "forward" into the future. This vision of the relationship between past, present and future realities lies at the heart of the popular narrative of modern industrial society and serves as the reason and excuse for co-opting natural resources for the purpose of exerting power and domination over the entire world. The neoliberal vision of the social, political and economic order provides the rationale for this view of human habitation on Earth—this worldview regarding the nature and purpose of the human story. Further, it provides the motive force for preferred activities and perspectives of individual actors relative to the overarching concept of human life on Earth.

Hence, modern society has evolved under a runaway, neoliberal free market economic system and has done so at the expense of a natural world of resources erroneously viewed as infinite and a human population that has succumbed to the belief that consuming industrialized products is not only equated with upwardly mobile economic status and happiness, but also represents a form of responsible citizenship. Once upon a time there was a fundamental emphasis on the inculcation of values associated with citizenship for active participation in a democratic union. However, this is no longer central to the Western narrative of democratic ideals (Cassell & Nelson, 2013). The neoliberal agenda demands allegiance to individualism, competition, and blind obedience to authority and redefines citizenship in terms of consumerism and as a manufactured form of patriotism, as in George W. Bush's post 9/11 declarative to the nation to "go shopping." Monbiot (2011) argues that neoliberalism views competition as "the defining characteristic of human relations" and any attempts to limit competition are directly contradictory to once held notions of citizen rights and liberty.

The transformation of western civilization away from social orientations aligned with the public good, valuing roles and responsibilities of citizenship as central to democracy and toward the notion of a society based upon a neoliberal approach to the privatization of almost all aspects of democratic participation has been relatively rapid, widespread and all consuming (see also Cassell & Nelson, 2013). A new state of social engagement has arisen amidst an overwhelming indoctrination of fear, terrorism and insecurity, wherein winners and losers are defined through economic, political, legal and educational systems aimed at further stacking and solidifying socio-economic stratification in favor of extant elites while, at the same time, providing persuasive rationales for the unquestioning, lock-step cross-generational reproduction of these increasingly hegemonic power structures (Cassell & Nelson, 2015). Social values and belief systems are being redirected toward an Orwellian version of a future that both requires and diligently creates a citizenry loyal to conformity, obedience, and compliance to a normalized political and economic state of never-ending war and the employment of institutional violence as the dominant strategy for solving human problems. In addition, this neoliberal "democratic" ideal is based on an increasingly quaint and arcane mythology lodged in the intellectual dicta of a nascent form of capitalistic market ideology first conceptualized by Adam Smith nearly three hundred years ago—a conceptual framework born of another time in history in which humanity lived on a drastically different Earth where environmental scarcity raised no present issues for an embryonic system of global trade and resource redistribution. Where better to inculcate future generations with this mythological vision and train them for living this dystopian dream than through the educational system? Where better to indoctrinate young people with belief systems and particular forms of knowledge that protects, elevates and serves those in power? Where better to prepare a future workforce to be docile, fearful, obedient, compliant, and willing to accept and perpetuate a social structure from which only a few will reap the benefits? Where better to engineer a future wherein the natural world is subjected to monetary coding void of any intrinsic or affective human meaning and value?

This worldview has dominated the hegemonic formulations underlying Western orthodoxy for centuries, however, the iteration active at this point in human history has significant potential for wreaking irreparable destruction upon the natural world, our only home. We are at a time in human history where a deepening divide across cultural, social, economic, and ecological aspects of the human experience are threatening not only the long term prospect of human inhabitation on Earth, but also the very diversity of life, the foundation upon which civilization has risen, and now appears to be headed for imminent mass species extinction and collective eco-suicide. It appears this extreme and radicalized agenda has gained incredible traction within the larger Western educational systems. Perhaps it is simply a function of the human genome to aspire to denuding and poisoning the environment to the point of self-induced genocide awash in our own waste stream.

The rapid expansion of corporate and private sector involvement in public schooling over the past thirty years has resulted in a particular vision of educational reform attuned to and intertwined with neoliberal capitalism, in promoting and designing a standardized curriculum and testing paradigms aimed at viewing students as both consumers and a future low wage labor force (Cassell & Nelson, 2013; Cassell & Nelson, 2015). This business minded orientation to schooling and education has become a key component of economic globalization and significantly impacted policies and practices within local and regional contexts in the administrative, theoretical and practice-oriented landscapes of US education. Over this time, curricula have become ever more narrow and focused on language arts and mathematics (at the expense of the natural and social sciences, arts and humanities, history, civics, and ecology in particular). In addition, the textbook/curriculum materials and testing industry has seen enormous growth during this period to the point where this form of top- down, content mastery-focused, career niche preparatory "educational reform" has become standardized, test-driven, and normalized.

The concentration of authority over control of goods can be applied to ever more restrictive policies governing teaching and learning practices in schools today. Corporate control over educational policies has resulted in narrowing curricular opportunities for both students and teachers, while at the same time expanding profit margins. The standardization reform movement has resulted in less local autonomy and emphasizes an education that promotes hyper-consumerism and selfish misuse of natural resources (Nelson & Coleman, 2012).

Long gone are broader publically minded visions of preparing students for citizenship in a democracy. They have been replaced with a narrowly focused, outcome oriented obsessive-compulsive urge to prepare students for a decadent world of ruthless competition, hyper-consumerism and militarized labor conditions rationalized as a way to enhance American students' ability to compete in a global lions' den. Virtually entirely absent—except for largely mindless nods in the direction of intellectually arid and superficial diversity oriented political correctness—is any substantive form of phenomenological curriculum or pedagogy directed toward leveraging students' places, people and lifeways, thus unlocking the intellectual and emotional intelligence of non-dominant cultural formulations found in the life experience of increasingly diverse student populations (Cassell, 2015; Jehangir, 2009; Slattery, 2013).

The social impact of these corporate-driven policies has wreaked havoc in communities across the country, in particular those in poverty-stricken urban neighborhoods, and has confined students, parents and communities within an economic system that values only the profiteering of the financial and industrial sector. As these same corporate interests have dramatically influenced democratic government policies and practices through unregulated lobbying, the forces that now control a nationally normalized education system have defined and relegated the public into patterns of passive acquiescence and outright servitude to corporate and militarized interests.

The Problems with Contemporary School Reform

Contemporary school reform and subsequent classroom practices do not appear to be taking human-environmental relationships and subsequent critical ecological issues seriously. As alluded to above, much of what passes as official curriculum today looks frighteningly narrow and mostly absent of context. Formal education has responded only minimally to the ecological crisis, and may actually be considered a direct contributor to the kind of knowledge production inherent in furthering environmental degradation, especially when it comes to addressing climate change. It is generally agreed that humans are having a major influence on the health and well being of all living things on Earth. Much of that influence, as noted above, has had dire consequences in the form of species extinction, over consumption of natural resources, and the dramatic rearranging of our planet's chemistry. Curriculum has long been perceived as official knowledge, knowledge "worth knowing" from reference points established by the culturally, politically and economically dominant groups atop the societal pyramid. It has typically been organized into discreet units of pre-packaged subject matter presented for the purpose of standardizing what all students should know and be able to do. Traditionalist or Essentialist conceptions of curriculum continue to dominate time and content in schools at the expense of meaningful engagement in the kind of actionable applied knowledge necessary to solve real-world problems such as helping to reimagine the human-environmental dynamic and work towards healing a planetary ecosystem literally on the brink. There appears little institutionalized sense of urgency and little consideration that humanity has been directly complicit in species extinctions and subsequent loss of biological habitat and diversity—and perhaps now faces the stark reality of its own impending extinction.

This should, however, not come as much of a surprise, as schools rarely have been charged with extending experiential learning beyond classroom walls, or bridging curricula and pedagogy across the messy expanse of an increasingly multicultural society. Contemporary schooling fails to engage students in mining the intellectual richness inherent in this expanding and diverse pool of community wealth and authentically taking on emerging socio-cultural issues from multiple perspectives, let alone the ecological crisis that threatens not only all of the Earth's ecosystems, but the very existence of humanity itself. What little attention emerging issues receive in schools—other than the largely "safe" politically correct flavor of the day approach— is typically relegated to disparate and minimally contextualized content, and too often embedded in discipline-based textbooks. Official knowledge is ultimately based upon political and economic ideologies that have enormous influence over what is taught and how it is taught in school classrooms (see Apple, 1993; Bednar, 2003; Spring (2004).

Although contemporaneous political and economic ideologies have historical-ly had a significant influence on school curriculum, what has remained largely un-changed is an organizational structure that artificially separates content disciplines and rarely offers opportunities for interdisciplinary and cross-cultural ways of know-ing necessary for addressing emerging and critical issues in an ever shrinking global world-system, even should they be considered politically appropriate and more in line with the circumstances of today's world than the increasingly outdated ideas carried forward from previous eras. Throughout the history of public education, curriculum has changed little in terms of disciplinary organization. Schools today, more than any other institution, are remarkably similar to how they were organized and how they looked 150 years ago. The emphasis has remained focused on preparing students for work that too often results in business related activities associated with unchecked extraction and/or harvesting of natural resources, resulting in widespread collateral ecological destruction and environmental deterioration. Rarely do students in school experience a full and inclusive phenomenologically oriented curriculum aimed at understanding alternative, even historically marginalized, ideas relevant to the search for solutions to the emerging complex social, cultural and environmental problems of today and the necessarily connected search for and identification of interdisciplin-ary and, increasingly, intercultural ways of knowing and acting. As a result, there are rarely opportunities for teachers and students to learn about human and non-human systems and how they inextricably interconnected (Cassell & Nelson, 2010).

Further, both social science and hard science textbooks and curricula routinely present the history of the world as an inexorable march away from what are often depicted as "noble savage" primitives mired in purely reactive and defensive forms of interaction with the world and toward more advanced and empowered forms of social organization and structure capable of manipulating the natural environment and bending it to the will of an all-powerful human intellect (e.g., the logical sequence of chapters and units in both elementary and secondary history textbooks tracking the patterns of intellectual and social development from the end of the medieval era through the Age of Reason and the rise of the Enlightenment movement in Western Europe). Today, school curricula are clearly at odds with what will be required as necessary skills and knowledge in the future, particularly those required for and focused on ameliorating environmental and social problems. In addition, students today are being prepared for jobs and skills that are no longer viable in a rapidly changing global economy, in a rapidly deteriorating planetary biosphere—the rationale for which is part of another period of human history, a worldview now hopelessly arcane and deleteriously out of step with the realities of life on planet Earth in the year 2016.

What constitutes educational dialogue today typically centers around politically charged issues like teacher unions, parental choice and school vouchers, charter schools, teacher retention, standardized testing, common core standards (i.e., a national curriculum), teacher accountability tied to student test scores, the teaching of evolution and/or creationism. The Business Roundtable (www.businessroundtable. org) has had enormous influence over the policy environment and the accompanying debate over educational policies in the United States. It is important to note that these policies and mandated practices are exported as components of foreign aid packages to countries around the world (see Cassell & Nelson, 2013, Spring, 2004) and represent the strong arm of US educational policy driven by free market enterprises and corporate for-profit interests. Rarely, if ever, are the underlying assumptions of global economic competition serving as the driving force behind school curricula and educational reform efforts questioned. There seems to be embedded in the current cultural ethic of education an assumption that competition is far and away the primary goal and determining factor over who wins and who loses. As a result, historically, little effort has been placed in school on cooperation and collaboration around shared goals and interests. Even less effort has been placed on examining a priori assumptions underlying the shape and texture of the social fabric and searching for new, possibly more efficacious, formulae relative to the ordering of society for the purpose of enhancing the health and sustainability of human life on Earth. Clearly, if human civilization is to survive the present global ecological crisis it will require a pan-national, pan-societal and pan-cultural cooperative effort never before seen in human history. What is desperately needed is a mechanism for creating habits of mind among the citizenry that lend themselves to successfully engaging in just this endeavor—habits of mind centered on an appreciation for the unique gifts, insights and cultural wealth of "the other".

The global reach of the most powerful energy corporations and their subsequent responsibility for the production and streaming of toxic waste within and through-out local communities is undeniable. Entire neighborhoods and surrounding regions, typically immersed in conditions of poverty, have become dumping grounds of dis-carded toxins. The domination of food production by transnational corporations has negatively transformed the ability of local communities to live sustainably while touting the benefits of globalization. These companies view the resultant natural and human habitat destruction as unintentional, but necessary collateral damage, as busi-ness as usual in the extraction of oil, coal, and other minerals associated with carbon based energy production and the industrialization and corporatization of agricul-tural food production. Because of their enormous influence over national and local politics through large scale lobbying efforts and the direct funding of local, state and national candidates for office, multinational energy companies such as Monsanto, Bayer, BASF, Cargill Exxon Mobile, Chevron, BP, Koch Industries and Massey Energy are essentially co-opting the ability of citizens everywhere to participate in decisions that have direct impact on their own lives and livelihoods. The idea of democracy as central to national decision-making has lost meaning and influence, to the degree that the very ideas embedded in democratic ideals are rapidly disappearing. Where in the formal school curricula are found the imperatives, the ethical questions and considerations necessary for responding to these hyper-capitalistic forces in morally critical and democratic ways? They are certainly not to be found in educational pro-gramming provided through corporate—school district partnerships through which funding and curricula are provided by the very businesses in whose interest it is to engage in unfettered mass-environmental destruction and which depend on a passive and quiescent citizenry content to stand by and watch Armageddon bear down as they drive to the shopping mall.

Educational Conundrums and a Pedagogy for Survival

What are the roles and responsibilities of educators in addressing issues and prob-lems associated with cultural and social causes of the ecological crises? In what ways are schools and curricula complicit in promoting and valuing behaviors that ulti-mately contribute to the exacerbation of the ecological crisis? As noted above, histor-ically, schools have rarely responded to cultural, social, political, economic and en-vironmental challenges as a function of its larger societal charge. Formalized school curricula are typically organized around discrete subject matter content areas, and contemporary schooling policies have narrowed the answer to the question, what does it mean to be well educated, to the regurgitation of specific answers to specif-ic canned questions, typically formulated at the most shallow end of the learning spectrum where the effectiveness of the exercise is calibrated on the basis of inauthentic decontextualized forms of formative and summative assessment. Clearly, this approach to learning is designed to measure the simple acquisition of data points and information—as opposed to the ability to use critical thinking skills to ponder aspects of the complex systems-based reality of life on Earth in search of optimum responses to open-ended questions. It is also clear that schooling policies, and teacher education policies associated with them, are aimed at maintaining an educational system driven by the most simple and controllable pedagogies. What purposes does this approach to education and educational policy serve? If indeed our global and local environments are in peril from runaway political and economic greed, where in society rests a moral and ethical stance toward responsibility for future generations? If education is considered a moral act, then where are the voices and subsequent actions from the profession that resonate powerfully regarding the moral responsibility for our children's future?

The authors aim to engage others in ways of thinking about teaching and learning that are rooted in interdisciplinary knowledge, in understanding interconnections and interrelationships—ways of thinking that are true to the actual structure of life of Earth, a structure grounded in the interrelationship of complex nested systems. We find it particularly troublesome that traditional school curriculum is organized in ways directly in conflict with this fundamental reality of life on Earth and how knowledge is actually utilized in the larger human endeavor. Human problems are multifaceted and complex. They are rarely solved through the lens of one discipline. They are addressed and solved in collaborative contexts wherein interdisciplinary ways of knowing are expressed through content specific analysis and cross content synthesis. The implications for such a paradigm shift are significant and require attention to larger moral and ethical purposes. Poverty, social justice, eco-justice, multiculturalism, and issues of educational equity are all essential aspects of a broader global cultural crisis that includes at its center human-environmental relationships resulting in worldwide ecological systems collapse.

The institution of formal education must become a dynamic mechanism through which potentially deleterious forms of hegemonic thinking are challenged as the life experience and cultural wisdom of previously marginalized populations are brought to the center and allowed to compete with historically dominant intellectual formulations in open, creative and contexualized learning spaces. As dire as the present situation is and the future possibilities portend, we believe that teacher educators and PK-12 classroom instructors taking action and deploying education as a tool through which to leverage multiple cultural and intellectual points of view in the service of generating new formulations appropriate to address the constantly changing context of life on Earth is a critical part of finding solutions and, further, is central to the true mission and ultimate purpose of education. That is, education must operate not solely as a vehicle for the dissemination of some consensus form of positivist preferred or "official" knowledge, it must be an engine for the creation of reality-based phenomenological forms of knowing tied to the actual settings and circumstances of people's lives. Over time, this form of knowledge might become a matter of general consensus, but only if it survives continual reassessment and reevaluation against the current circumstances of human life on Earth. Sadly, formal education in the West has not tended to function in this way. As a result, it has not been able to address ecological crises in any meaningful way and, in fact, works to reproduce a kind of knowledge and resultant behaviors that actually exacerbate human-environmental problems (Cassell & Nelson, 2015; Nelson & Cassell, 2012).

Nelson and Coleman (2012) argue that efforts to contextualize school curricula within a framework of human-environmental relationships has been relegated to a politically radical status, something to only consider outside the parameters of what is considered official and legitimized knowledge. Contemporary political agendas have gained widespread and "ever-increasing leverage over public discourse about what it means to be well educated and has stifled attempts to legitimize counter efforts aimed at broadening public school curricula to be more socially, culturally, ethnically, morally, and environmentally responsive" (pg. 155).

What is required is a major paradigm shift and new ways of thinking about our shared existence with myriad life forms on Earth. This will necessitate a radical change in the way education is conceived, redirecting its' purpose toward a more phenomenological, cross-cultural and reconstructionist orientation aimed at equipping younger generations, who will inherit a world falling into ecological collapse, with the tools, skills and knowledge essential for ameliorating a politically, socially, culturally and economically driven planetary crisis in an authentically impactful way.

Despite short and long-term efforts of those involved in environmental education to bring the field into the center of the national discourse over the past 40 years, it remains mired in a zone of curricular purgatory. At best, it is typically seen as a marginal field of study. Although Orr's (1992) now famous proclamation remains a point of solidarity for environmental education professionals, the notion that *all* education is environmental education has yet to become rooted in either the professional discourse or policy-making activities occurring across the mainstream of the education and schooling establishment—much less the broader spectrum of socio-cultural debate. All too often, environmental education's approach to seeking mainstream acceptance has failed to gain traction. Even now, more than fifty years after Rachel Carson painted a gripping and grizzly picture of the ways in which humans are poisoning not only natural ecosystems, but their own lives within those ecosystems, environmental education remains marginalized and sequestered into the somewhat isolated intellectual space associated with *environmentalism*, which is identified with a set of ideas fundamentally distasteful to wide swatches of America's power elites as well as large segments of the general public.

Hence, environmental education continues to occupy the unenviable position of being easily dismissed as "politically correct" extremism. The question before us, given the array of seemingly intractable socio-cultural issues revolving around humanity's predilection for destructive behavior towards the environment, is how to pull environmental education out of the hinterland of the intellectual margins of the American scene and into the center of public discourse; how to bring it to a wider audience and free it from its' lock-step attachment to the environmentalism niche and the separation from the intellectual mainstream this has engendered.

It is our contention that the question of the movement of environmental education and learning into a more central place within the national conversation is of critical importance to our society and culture at this given point in time and, further, that the predicate considerations related to the host of potentially deadly environmental crises discussed above lie at the center of why this question is so terribly urgent. Elsewhere the authors have proposed the concept of a *pedagogy for survival* (Nelson & Cassell, 2012). This concept posits that formal educational institutions should serve to challenge and disrupt deleterious and potentially catastrophic hegemonic formulations regarding the nature of knowledge, knowing and preferred modes of applying knowledge in the pursuit of social and cultural goals and objectives in the post-industrial West. The specific frame of reference for this discussion is the need to reformulate the increasingly unsustainable and ruinous patterns of human habitation on Earth. The simple and compelling reality is that all other considerations regarding the form and substance of the educational establishment and, for that matter, all other forms of human activity, are rendered moot by way of irrelevance if humanity destroys the planet and, by so doing, renders itself extinct.

Of key importance in considering the issue of formal education's role relative to potentially dangerous hegemonic propositions which serve as the basis for patterns of socio-cultural dominance is the concept of habitus, that is, deeply embedded socio-culturally delineated triggers for decisions related to making the choices that guide actions at a subliminal level of consciousness. If we are to make emotionally healthier, more socially just and culturally expansive choices, then the shackles that bind us to arcane, decontextualized forms of habitus must be broken. As Bourdieu noted, formal education often serves as a mechanism for the blind reproduction of hegemonic formulations through the unquestioning propounding of these formulations through curriculum and pedagogy deeply rooted in prevailing forms of habitus (in Nash, 1990). However, it is the contention of the authors that formal education can be structured so as to play a different role and break this deleterious cycle—actually functioning to move society and culture forward in ways that intelligently respond to the adjustments and accommodations of the dynamic, open and complex system structure extant in the socio-cultural polity itself and in the broader array of natural and human systems in which it is nested (Cassell & Nelson, 2010; Cassell & Nelson, 2015; Haggis, 2008). It is here that we find a powerful rationalization for bringing environmental education and learning front and center in the increasingly fervent debates about our planetary future.

The authors approach questions of the structure and purpose of education from the perspective of systems thinking and systems theory—though our approach stems from what may be a wider perspective than environmental learning and/or sustainability education per se, these fields are nonetheless inexorably tied to, and derive their ultimate intellectual power from, the key principles of systems structures and operations. More broadly, this perspective carries with it a theoretical framework based upon the proposition that the human experience is one situated in the context of the operation of dynamic, open and complex systems. As these systems are dynamic, they involve a large number of components that interact dynamically at a local level. These interactions are non-linear, that is, they involve feedback loops that continually adjust and modify both the parts of the system and the overall system itself. Dynamic systems are open systems; therefore, interactions can also affect the boundaries of the system. That is, the membership of the system can be changed, enlarged or constricted as a result of the change process initiated by the interactions of its parts. An open-system architecture also means that these interactions result in changes that have effects beyond the system. These interactions are multiple and they are connected in complex, multifaceted ways and at numerous levels of operability. It is the multiplicity of the interactions through time that produces effects and changes at the level of system components and the overall system structures themselves. In a true systems configuration, causality cannot be reduced to single or limited number of factors; all the factors operate upon each other. In such an operative dynamic, there will typically not be one overarching causative factor responsible for effecting changes and adjustments in system configuration or functionality (Haggis, 2008).

As Caine (2004) has pointed out, the ultimate applicability of this approach to a consideration of the function and potential of formal education with regard to the disruption of deleterious hegemonic formulations rests squarely on the realization that human beings are themselves systems. Further, the human animal is a complex system in which embedded elements are interrelated and their interactions create changes in the elements as well as the entire system of which they are part. It is characterized by embedded interconnectedness and the interactions within it are nonlinear; they create effects in numerous directions and not all such effects are expected or anticipated. So, there is an element of the unpredictable and unanticipated with regard to the effects / results of these interactions. The operative force here is, then, a dynamic embedded nonlinear interconnectedness. The inherent interconnectedness of the physical features and processes of learning in the human body are now well known and have been scientifically demonstrated (Jensen, 2005). The human body and mind form a system in which everything is interconnected to everything else and everything influences everything else in multiple ways.

As Greenwood (2010) has observed, the structural realities discussed above exist in awkward juxtaposition to a deeply engrained form of practice in public schools (and, for that matter, teacher education programs) that has resulted in a view of learning as a mechanistic process undertaken in a factory-like, highly bureaucratized setting and structure. It follows then, that to produce positive changes in the institution of formal education that can equip it to address critical questions about our socio-cultural destiny, we need to change the historically dominant view of how people learn.

This must be brought into greater alignment with the systems structure that forms the basis for the nature of life itself and the processes of life in the natural world in which humanity exists (Caine, 2004; Cassell & Nelson, 2010).

Bowers' (2008; 2009; 2010) has set the problem relative to education and hegemonic thinking in historical perspective and stark relief through his examination of the conundrum presented by the epistemological framework of the Euro-centric cultural matrix assembled in the wake of The Age of Reason in support of the rise of the global capitalist economic model, the industrialized economies of the West and the patterns of wealth accumulation and resource utilization resulting from them. Bowers asserts that the present condition in which humanity finds itself relative to the threats inherent in its "global footprint" is the result of thought patterns and worldviews inherited from another time. As discussed above, we are, to a great extent, held hostage by outmoded intellectual paradigms increasingly out of place in the time and world in which we now live, but which still wield tremendous power over our minds, our ambitions and our conception of "Earth as home".

Bowers refers to this epistemological trap as "double bind thinking" and states that it leads us to carry forward the outmoded ideas of earlier thinkers who were free of the constraints imposed on our generation by an ever more urgent crisis rooted in a growing number and scope of demands being placed on a world beset by a rapidly increasing scarcity of resources. He advances the idea that what is called for in a world of critical scarcity is a new root metaphor—one based on ecological forms of thinking. This ecological root metaphor is most properly understood as a systems theory-based worldview. Bowers points to the need for a new linguistic tradition in support of this worldview, a vocabulary for life lived within, and as part of, a complex global system as opposed to life lived overseeing and tinkering with a global machine.

Bowers' position speaks to the need to rethink and restructure the form and substance of institutionalized learning and formal education in light of the very real risks inherent in adhering to dangerously arcane hegemonic propositions inherited from previous generations. Formal education should be recast so as to lessen the current emphasis on abstract, disembodied theoretical knowledge and the search for pat answers to canned formulaic inquiries and move toward more meaningful and authentic forms of knowing aligned with the systems architecture of the world so desperately in need of balance, relief and support.

Goleman (2009) proposes a schematic relative to knowledge construction and utilization that runs in line with the systems-based approach described by Haggis, Caine and Bowers. He advocates for the development of a new form of collective intelligence as a solution to the cognitive dissonance described above. The depth of awareness needed to see and comprehend the risks and threats all around us is beyond the cognitive capabilities of any single individual—or, for that matter, culture. Hence, we must develop and employ a powerful form of shared intelligence in order to successfully address the new threats inherent in the multifaceted global crises now pressing down upon us. In a very real sense, we need a pan-societal and pan-cultural form of distributed intelligence wherein the implicit cognitive load is spread among a wide array of persons with highly varied experiences, perceptions, skill sets, native intelligences and cognitive approaches to learning (i.e., "learning styles"). What is called for, then, is a system of distributed cognition and memory that allows us to comprehend more fully the totality of the complex multi-tiered array of systems that is, in fact, the world we live in. Goleman calls this *"swarm intelligence".*

It is within this context that he finds great potential in the Internet as a platform through which people share experiences, concerns and insights. However, there are also obvious and critically important implications for curricular structures and instructional paradigms in public education. Goleman's observations clearly point to an urgent need to redirect the nature and shape of formal education in order to rediscover and utilize a wider array of intelligences and learning styles in an effort to leverage our distributed intelligence in order to create more holistic visions of the world in which we live.

This effort will, out of necessity, require restructuring the learning environment in schools so as to build an awareness of coherence, community, shared fates and destinies, and the inherent value of every person in the form of that person's unique talents and abilities—the unique contributions that every person can make in asking important questions and solving complex problems. In essence, we need to create a learning environment in which "the swarm" can begin to take shape.

Here then lies the critical utility and inherent power of environmental learning, sustainability education and the ecological/systems metaphor that undergirds them. It rests upon the holistic, multifaceted, multidirectional non-linear form of interaction (both physical and intellectual/cognitive) that lies at the true center of life on Earth and provides an actionable framework for bringing an expansive wealth of individual talents together in powerful community contexts. This holds forth the possibility of developing more creative, useful and expansive forms of understanding relative to the natural forces that dictate the parameters of human habitation on Earth and presents the opportunity to more effectively address the potentially catastrophic planetary circumstances with which we are now confronted.

However, to make this a reality, the form and substance of formal educational institutions, beginning with PK-12 public education, must be fundamentally reimagined and recalibrated accordingly. We must create relevant an authentic learning contexts that leverage and act upon the naturally occurring system dynamics that drive the function of the human mind. We need new paradigms in educational theory and practice that take us back to fundamental systems-based forms of learning and knowing. Formal education must, then, be organizationally structured around the actual reality of the role of learning in human life. This requires that it be contextualized and tied to the life experiences of learners in ways that give it relevance, meaning and power. Our lives are not lived in artificial, compartmentalized niches. School structure and organization must be torn apart and rebuilt to reflect the actual context of our lives—that is, multi-dimensional, interconnected, open-ended, cognitively "messy" and—at its' core—quintessentially phenomenological in nature. Classes based on artificially discrete subject disciplines in which students are asked to find pre-determined answers to closed-ended questions which carefully delineate pre-set alternatives must give way to block-scheduled classes of longer duration based on the interdisciplinary cross-fertilization of subject matter in which cross-departmental team teaching leads students in exercises of open-ended critical inquiry designed to ask relevant and life-centered questions, identify issues and delineate problem scenarios in complex, open-ended academic activities that stress authentic links to the lived experiences of students and employs the natural learning modalities of the human mind such as trial and error. Students must be supported in this by being urged to take risks, fail, and learn from their failures, hone their thinking and academic products and recalibrate their approach to problem solving (Sternberg, 1985a; 1985b; 1988).

It follows, then, that learning activities must be meaningfully linked to the living contexts in which they occur; they must make use of topics, issues and inquiries intimately connected with and drawn from the lives of learners. The concept of interconnectedness must come alive in curricular structures that endeavor to build, adapt and transfer knowledge as opposed to simply memorize disembodied content so as to undergo inauthentic assessment exercises. This means that curricular structures must be more flexible and open-ended and should overtly leverage a sense of community and locality. Schools need to build content around authentic learning activities that mirror the communities, localities and life circumstances just beyond their walls so that learning ceases to be an artificial and physically sequestered exercise unaligned with the broader contexts of human life (Wenger, 1998).

Pedagogical practice will have to be aligned with structural and curricular reforms to reinforce an authentic contextual approach that works against the lockstep cross-generational reproduction of anti-system oriented hegemonic formulations. Pedagogy that sets learning in highly interactive situations in which socio-cultural and cognitive conflict forces students to examine themselves and the a priori assumptions upon which their expectations relative to themselves and others are based will be critical in the effort to destabilize the foundations of "taken for granted" socio-cultural formulations which drive the operation of endlessly recycled habitus (Jehangir, 2009). In this way the process of learning becomes one of socio-cultural reformulation in which learners develop and hone identities of competence and participation. Learning becomes genuinely transformational—a process of "becoming" in which learners can explore their relationship with one another and with the general landscape of their lives (Wenger, 1998).

Group work pedagogies that utilize learning community settings can be invaluable in challenging the assumptions that support the unquestioning perpetuation of normative approaches to the relationship between human habitation and the natural environment. Collaborative learning venues in which students work together on open-ended, complex tasks that involve problem solving activities in which there is no one "right" answer and work is done through relationships of genuinely reciprocal interdependence in which the work, learning styles and intelligences of any given student are necessary ingredients for the completion of their colleagues' work can leverage cognitive dissonance in operative venues that provide real, viscerally felt experiences in which students function as vital nodes in complex system structures (Cohen, 1986; Cohen, Lotan, & Leechor,1989).

Such pedagogical approaches are, then, ultimately reliant upon and can serve to activate the foundational forces of distributed cognitive processing which are the essential predicates to the unleashing of the full power of the systems architecture embedded in environmental learning and sustainability education. In doing so these instructional techniques can powerfully reinforce the interdependent system-based realities of the human mind and life on Earth. Deeply embedded fundamental cognitive processing routines such as transactive memory systems and processes of cognitive load spreading can serve to channel structured social interaction built around active verbal engagement and socio-cognitive conflict throughout collaborative work groups in ways that reinforce key group systems dynamics such as holism, interdependence and complementarity (Kirschner, Paas and Kirschner, 2009; Michinov & Michinov, 2009).

Pedagogies rooted in systems-based instructional modalities are also powerful venues through which weak tie social capital networks can serve as conduits through which human capital can flow between strong tie relational clusters, facilitating the flow of potentially transformational information across more distant parts of the social structure (Granovetter, 1983). These bridging weak ties serve to connect individuals who are significantly different from one another in ways that stimulate mutual understanding and cognitive flexibility. This will serve to create viscerally felt experiences of systems-based operations in the social structure that counteract disjunctive hegemonic formulations.

The authors assert that curricular and pedagogical formulations structured as described above can create the foundations for a construct of learning and knowing which can serve to reawaken humanity's viscerally felt understanding of the operation and structure of both social and natural systems and the critical interrelatedness of the two. It is this understanding and the accompanying appreciation of the opportunities and limitations inherent in the overarching systems architecture of the world which can free us from behaviors that are artificially suppressing our ability to move Western society forward in a direction that can optimize the relationship between the natural and human systems in which we live. Environmental learning and sustainability education can play a potentially powerful and central role in this scenario. Both these disciplines are based on an in-depth analysis of the systems architecture of local and global ecosystems and the biosphere. They engage in these analyses from multiple perspectives—scientific, economic, political, cultural, social and policy-making. Further, they seek to understand and optimize the forms in which human habitation relates to and impacts the natural environment of the planet. In so doing they can break out of the environmentalism niche in which they are currently encased and move away from the margins of intellectual endeavor in contemporary America.

The Connective Tissue of Solutions to Contemporary Crises: Systems Thinking, Social Justice, Eco-Justice and Cross-Cultural Problem Solving

It is in the context of unleashing Goleman's swarm intelligence through authentic forms of systems-based human interaction that the construct of eco-justice in the twenty-first century must be understood. It is, in point of fact, closely related to reconceptualized twenty-first century formulations for multiculturalism and multicultural education and to a fundamentally utilitarian construct of social justice that emerges from the energizing of these formulations through the impetus of systems thinking. The use of institutional educational venues as laboratories for unleashing these phenomena can provide an operational framework through which they can take shape and work to influence the future course of human history.

Marouli (2002) points out what the authors have noted above, culture plays a critical role in shaping the relationship between humans and the natural world as well as the outcomes of that relationship. We understand the world around us through the cultural manifestation of value systems. Indeed, our perceptions of the world are culturally situated. She posits that, in a very real sense, the current crisis of environmental degradation is a crisis in the value systems of the societies in which it occurs. Hence, the cultural contexts in which the human-environment relationship is forged are critical in coming to understand that relationship and addressing potentially deleterious deformations within it.

Marouli (2002) also points out that environmental education is increasingly focused on these issues through the emergence of sustainability education and its emphasis on finding formulae through which we can "live for sustainability" in an effort to lessen the pressure humans are placing on an increasingly fragile planetary environment. Marouli asserts that sustainable living as an implemented policy and, as McLaren and Ryoo (2012) term it, a "lifeway" must necessarily deal with the relationships among and between equity, economics and the environment as well as the actual policy actions and plans to make required changes in human habitation patterns. This effort carries along with it, by necessity, a host of social, cultural, political and economic considerations. Hence, the socio-cultural dimensions of environmental issues and problems are increasingly necessary components of environmental education and any possible solutions it helps to facilitate. It follows, then, that the overall approach taken here should be holistic, interdisciplinary and squarely aimed at problem solving within the framework of community structures and needs (Marouli, 2002). This necessarily begs the question as to the role of cross-cultural thinking and policy making in addressing the world's current environmental crisis.

Marouli goes on to raise another aspect of the connection between culture and the environment which links directly to the status, nature and use of multiculturalism and multicultural education in society generally as well as in the search for more workable and sustainable forms of human life on Earth. The increasing globalization of life on the planet and the increasingly pan-global nature and ever-expanding complexity of our current environmental crises has created an urgent need to communicate across national, social and cultural boundaries in order to find strategies through which we may approach these intractable problems. In point of fact, as alluded to above, any possible solutions to our environmental problems lie beyond any one person's or group's local reality or intellectual quiver. We are increasingly called upon to work together in the search for answers and, therefore, cross-cultural communications—as well as cross-cultural intellectual fertilization—become critically necessary (Marouli, 2002).

Marouli (2002) builds on Goleman's concept of a distributed "swarm" intelligence and provides an illuminating backdrop for the points of connection between social justice, eco-justice and emerging forms of multiculturalism rooted in systems thinking in arguing that, in part, multiculturalism and multicultural education are key components in environmental learning because building bridges across cultures in the search for solutions to environmental problems is not just a matter of diversity for its own sake. Rather, it is a matter of broadening the array of insights, perspectives, information, skill sets, problem solving approaches and sensibilities regarding human-environmental interrelationships that can be brought to bear on the problems at hand. All cultures have their own unique reciprocal relationships with the natural world, distinct from those of other cultures (Marouli, 2002; McLaren & Ryoo, 2012). These employ their own array of underlying cognitive pathways to knowledge and knowing and cognitive skill sets for applying them in the effort to know and successfully navigate the world (Gardner, 1993; McLaren & Ryoo, 2012). Any one or combination of them could yield invaluable strategies for addressing the nature of human life on Earth. Therefore, the inclusion of as many as possible into a truly open intellectual combine so as to provide a powerful forum of ideas on which to draw is a matter of ever-increasing urgency.

It follows, then, that environmental learning, multiculturalism and utilitarian constructs of social justice based on the need to provide seats at the societal table for as many different "lifeways" as possible are, in fact, connected at the hip. These fields of inquiry cross-fertilize each other at multiple levels and in multiple ways. However, perhaps the closest intellectual linkage between them is the structural concept of systems theory which, as noted above, lies at the very heart of human learning and cognition generally and, additionally, informs the core principals of environmental learning, sustainability education and cosmopolitan forms of multiculturalism. In addition, systems theory and ecological thinking speak to the question of how these key phenomena relate to the inculcation of social justice into the base habitus of human culture and society. This, in turn, can powerfully inform forward-looking concepts of eco-justice as an aspect of and tool for the creation of a truly empowering and pervasive form of social justice—one that can create a reliable basis for more sustainable forms of human life on Earth by creating more equitable forms of intellectual distribution and recognition between and among different forms of socio-cultural understanding.

The work of Marouli, Goleman and Gardner all point toward the utility, if not the absolute necessity, for a systemic form of inclusion in the intellectual life of humanity on Earth. There is no one perfect, all-encompassing way of knowing or understanding. The solution of complex global problems through the application of various socio-cultural formulations requires the marshaling of vast and multi-faceted intellectual forces. In addition, cultural perspectives outside those that gave rise to the problems in the first place have potentially immense value in the effort to address the deformative and potentially deleterious aspects of the originating socio-cultural constructs.

The vitally important role of involving diverse cultural constructs in addressing some of the most pressing problems of the twenty-first century mandates the establishment of conduits between cultures through which information and knowledge can flow in multiple directions. Approaching intercultural connections and exchanges from a systems perspective that reflects the actual structure of the human animal and the physical world in which it lives serves to enliven and empower all available resources at our disposal in addressing the problems with which we are confronted. It also frees us from the inherently limiting influence of segregated and exclusive constructs of individual and group identity. This opens the possibility of basing the relations of different cultural groups on a more realistic sense of identity—such as that propounded by Wenger (1998) in his description of what he terms communities of practice. Identity, as understood in this context, is viewed as a feature of integrated and ever-changing communities of shared endeavor. It is not a set-piece product of singular classification processes. Rather, it is a multifaceted, dynamic and evolving feature of humans engaged in processes of learning and acting that work to change the person as separate facets of identity (i.e., identities of knowledge and identities of participation) evolve concomitantly so as to impact one another and create new ways of being human in community—in concert with others. Each person undergoing such a process will necessarily impact the development of the persons with whom they interact and, ultimately, the form and substance of the communities of which they are part (Wenger, 1998).

This is, then, the connective tissue that ties the teaching practices described above as the core elements of what we have termed a *pedagogy for survival* to new forms of multicultural awareness and exchange and, through these, to a utilitarian view of social justice in which educational venues are purposely employed to bring forth a wide variety of cultural perspectives and "lifeways" so as to create a truly powerful form of socio-cultural pluralism which can be put to use in an effort to assess, reconsider and adjust or discard—as necessary—the taken-for-granted orthodoxies that serve to suppress the emergence of potentially more appropriate and powerful heterodoxies. Hence, social justice is not a matter of altruism, charity or even compassion—it is, ultimately, a matter of enlightened self-interest capable of and invested in recognizing the value and power in "the other" and the insights of cultural formulations other than one's own cultural heritage so as to attain one's own personal growth and enhanced empowerment (Cassell, 2015; Cassell & Nelson, 2015). Viewed against this background, eco-justice is properly understood as a variant form of utilitarian social justice through which multiple cultural perspectives are brought to bear in conceiving and addressing the environmental crises with which we are now confronted.

Humanity has arrived at a critical juncture in its history. Orthodoxies that do nothing more than support the continuation of the status quo so as to protect the position of societal elites must be challenged before it is quite literally too late. We need "all hands on deck" at this point in time, we must move beyond the constraints inherent in constructs of "preferred" or "official" knowledge. The future of the human family depends upon its ability to recognize that it is, indeed, a family. We must make classrooms laboratories in which ideas are created, tested and utilized in an effort to continuously reconceptualize the form and substance of the mechanisms through which we interact with the ever-changing natural environment of the planet.

❖ ❖ ❖

Thomas Nelson, Ph.D. is an Associate Professor in the Gladys L. Benerd School of Education, University of Pacific, where he coordinates the doctoral program in the Department of Curriculum and Instruction. He teaches doctoral courses in critical curriculum studies, teacher education and qualitative research methods. His areas of research include place-based learning, sustainability and environmental education, interdisciplinary ways of knowing, educational policy analysis, and social reconstructionist orientations to teaching and learning.

John A. Cassell, Ed.D. is a Visiting Assistant Professor at the Gladys L. Benerd School of Education, University of the Pacific. His research interests include connections between pedagogical practice and social justice oriented habits of mind, the social foundations of American public education, systems thinking in curriculum development and the role and purpose of public education in twenty-first century America.

References

Apple, M. W. (1993). The politics of official knowledge: Does a national curriculum make sense? *Teachers College Record*, 95(2), 222-241.

Bednar, C. S. (2003). *Transforming the dream: Ecologism and the shaping of an alternative American vision*. Albany NY: State University of New York Press.

Bowers, C. A. (2008). *The linguistic colonization of the present by the past*. Retrieved from http://cabowers.net/CAPress.php.

Bowers, C. A. (2009). *Educational reforms that foster ecological intelligence*. Retrieved from http://cabowers.net/CAPress.php.

Bowers, C. A. (2010). *Reflections on teaching the course "Curriculum Reform in an Era of Global Warming."* Retrieved from http://cabowers.net/CAPress.php.

Caine, G. (2004). *Living systems theory and the systemic transformation of education*. Paper presented at the Annual Meeting of the American Educational Research Association, San Diego, CA.

Cassell, J. A. (2015). Long and winding roads: Constructs for social justice and the nature of American public education. In J. C. Richards, & K. Zenkov (Eds.), *Social justice, the Common Core and closing the instructional gap: Empowering diverse learners and their teachers* (pp. 3-35). Charlotte, N.C.: Information Age Publishing.

Cassell, J. A., & Nelson, T. (2010). Visions lost and dreams forgotten: Environmental education, systems thinking, and possible futures in American public schools. *Teacher Education Quarterly*, 37(4), 179–197.

Cassell, J. A., & Nelson, T. (2013). Exposing the effects of the "Invisible Hand" of the neoliberal agenda on institutionalized education and the process of sociocultural reproduction. *Interchange: A Quarterly Review of Education*, 43(3), 245–264.

Cassell, J. A., & Nelson, T. (2015). Epistemology and apostasy: The role of education in times of neoliberal hegemony. In M. Abendroth, & B. J. Porfilio (Eds.), *Understanding Neoliberal Rule in K-12 Schools: Educational Fronts for Local and Global Justice* (Vol. I, pp. 335-354). Charlotte, N.C.: Information Age Publishing.

Cohen, E.G. (1986). *Designing groupwork: Strategies for the heterogeneous classroom*. New York: Teachers' College Press.

Cohen, E.G., Lotan, R.A., & Leechor, C. (1989). Can classrooms learn? *Sociology of Education*, 62, 75-94.

Gardner, H. (1993). *Frames of mind: The theory of multiple intelligences*. 10th anniversary edition. New York: Basic Books.

Goleman, D. (2009). *Ecological intelligence: How knowing the hidden impacts of what we buy can change everything*. New York, NY: Broadway Books.

Granovetter, M.(1983). The strength of weak ties: A network theory revisited. *Sociological Theory*, 1(1), 201-233.

Greenwood, D. (2010). A critical analysis of sustainability education in schooling's bureaucracy: Barriers and small openings in teacher education. *Teacher Education Quarterly*, 37(4), 139-154.

Haggis, T. (2008). "Knowledge must be contextual": Some possible implications of complexity and dynamic systems theories for educational research. *Educational Philosophy and Theory*, 40(1), 158–176.

Jehangir, R. (2009). Cultivating voice: First-generation students seek full academic citizenship in multicultural learning communities. *Innovative Higher Education*, 34(4), 33–49.

Jensen, E. (2005). *Teaching with the brain in mind* (2nd ed.). Alexandria, VA: Association for Supervision and Curriculum Development.

Kirschner, F., Paas, F., & Kirschner, P. (2009). A cognitive load approach to collaborative learning: United brains for complex tasks. *Educational Psychology Review*, 21(1), 31–42.

Marouli, C. (2002). Multicultural environmental education: Theory and practice. *Canadian Journal of Environmental Education*, 7(1), 28–42.

McLaren, P., & Ryoo, J. J. (2012). Revolutionary critical pedagogy against capitalist multicultural education. In H. K. Wright, M. Singh, & R. Race (Eds.), *Precarious international multicultural education: Hegemony, dissent and rising alternatives* (pp. 61–81). Rotterdam, Netherlands: Sense.

McKibben, B. (2011). *Eaarth: Making life on a tough and new planet.* New York: St. Martin's Griffin.

Michinov, N., & Michinov, E. (2009). Investigating the relationship between transactive memory and performance in collaborative learning. *Learning and Instruction,* 19(1), 43–54.

Monbiot, G. (2016). Neoliberalism: The ideology at the root of all our problems. *The Guardian US.* (April 19, 2016)

Nash, R. (1990). Bourdieu on education and social and cultural reproduction. *British Journal of Sociology of Education.* 11(4), 431–447.

Nelson, T., & Cassell, J. A. (2012). Pedagogy for survival: An educational response to the ecological crisis. In A. Wals, & P. Corcoran (Eds.), *Learning for sustainability in times of accelerating change* (pp. 63–75). Wageningen, Netherlands: Wageningen Academic.

Nelson, T., & Coleman, C. (2012). Human-environmental relationships as curriculum context: An interdisciplinary inquiry. In J. Lee, & R. Oxford (Eds.), *Transforming eco-education for the 21st century* (pp. 153–170). Charlotte, NC: Information Age.

Orr, D. W. (1992). *Ecological literacy: Education and the transition to a postmodern world.* Albany, NY: State University of New York Press.

Slattery, P. (2013). *Curriculum development in the postmodern era: Teaching and learning in an age of accountability* (ed.). New York, NY: Routledge.

Spring, J. H. (2004). *How educational ideologies are shaping global society: Intergovernmental organizations, NGO's, and the decline of the nation-state.* New York: Routledge.

Sternberg, R.J. (1985a). Teaching critical thinking, part I: Are we making critic al mistakes? *Phi Delta Kappan,* 67(3), 194-198.

Sternberg, R.J. (1985b). Teaching critical thinking, part II: Possible Solutions. *Phi Delta Kappan,* 67(4), 277-280.

Sternberg, R.J. (1988). *The triachic mind: A new theory of human intelligence.* New York: Viking Penguin.

Wenger, E. (1998). *Communities of practice: Learning, meaning and identity.* New York: Cambridge University Press.

Relational thinking in the Humanities and Social Sciences:
The educational dimension of eco-justice

Joseph Progler

As a process of discovering interconnections among what appear to be isolated entities and events, relational thinking is a way of knowing rather than an academic outcome; it is a means to an end, not an end in itself. Because it encourages students to move beyond seeing the world only through their own individual experiences or the lenses of specific academic disciplines, relational thinking can play an important role in higher education, in particular toward creating an interdisciplinary sensibility and fostering ecological and social justice.

As currently conceived, university curricula offer separate and distinct courses loosely bound together as majors and disciplines. By completing an array of courses within a major, students might achieve some understanding of a discipline. While descriptions of curricula allude to a way of making collective sense of these disparate entities, little attention is paid to how one course relates to another, beyond chronology, or how majors relate to student lives and their world, beyond getting jobs. Academic disciplines are often fragmented into sub-disciplines, each proceeding along a further narrowing trajectory. Professors trained in the disciplines reinforce these structures in their teaching.

Despite the fragmentary nature of academia, the role of relational thinking is implied by university mission statements. An administrative agenda promoting work-related skills, such as problem-posing and problem-solving, global awareness and teamwork, suggests opportunities for relational thinking. To pose problems, it helps to realize that many do not have immediate causes once their broader temporal and spatial implications are taken into consideration. To recommend that students become globally aware necessitates recognizing how the global and local might be intertwined. Being able to work as a team involves self-reflection, especially when team members come from different backgrounds and are unfamiliar with each other's cultural values and social mores and those of the societies and institutions within which they work.

Relational thinking is influential in several fields. Progress is being achieved in teaching mathematics to children, as reported by Carpenter, et al, (2005), who emphasize the equal sign as the key to understanding how numbers relate, while Stephens (2006) adds that relational thinking helps to develop children's mathematical understanding beyond computational skills. Arguing in favor of a relationship-centered economy, Schulter and Lee (1993) suggest that the ecological and personal ills of industrialized society can be healed through developing "relational markets." For the social services and helping professions, Burnside and Baker (2004) observe that prosecuting crime by relating the needs and rights of perpetrators to those of victims may bring about a more compassionate law enforcement, whereas Papadatou (2009) reflects on the importance of "relational care" in grief counseling. In philosophy, Didier (2012) insists that moving more toward recognizing relationships can deepen our perception of reality; research in physics and biology is proceeding along similar lines (Kineman 2011, Rosen 1991). A "relational turn" in geography, noted by Jones (2009), is facilitating an open-ended approach to understanding how people live and move within a space. Working on animal intelligence, behavioral scientists have discovered that crows exhibit an advanced relational awareness (Smirnova, et al, 2015). Computer science has notable features of relational thinking, such as in database programming and in the use of structured query language (Date 2011). Jazz musicians rely on a form of relational thinking that arranges performances around the relationship of melodies and chords to one another over time, enabling what Berliner (1994) calls the "infinite art of improvisation," while Keil (1995), Progler (1995) and Alen (1995) found that "participatory discrepancies," which occur as musicians spontaneously re-align themselves relative to one another, drive rhythm.

Academic disciplines among the social sciences and humanities, such as sociology, cultural anthropology and comparative religion, are predisposed toward relational thinking. By studying urban life and complex industrialized societies, sociology deepens our understanding of how individuals and groups are interdependent. Cultural anthropology has found that tribal peoples live in a relational world intimately informed by their ecosystems. Comparative religion provides insights into senses of human belongingness by looking at intrapersonal relationships and ethics.

With the preceding sketch in mind, this chapter will suggest contributions that the social sciences and humanities can make toward developing relational thinking. Drawing upon the author's experience teaching at an international university in Japan, the chapter illustrates relational thinking as an approach to teaching sociology, cultural anthropology and comparative religion. The arrangement of these courses within an interdisciplinary liberal arts framework has enabled a move from the conventional approach of delivering introductory content intended to prepare students for entering a major toward an outlook that provides the fundamentals of each discipline while shifting the focus to encourage forms of relational thinking.

Sociology: Developing Integrated Problem Posing and Problem Solving Skills

Sociology grows out of, and has direct pertinence to, the experience of living in complex urban industrialized societies. Due to the widespread reach of that society, among the three approaches illustrated in this chapter sociology is the most relevant to the lives of university students. At its core, sociology asks a two-fold relational question: What does it mean to live in groups and how does this impact the way we think and act? In reflecting on this question, sociology begins by positing the individual not as an autonomous entity but rather as a node in a web of relationships. Learning to recognize oneself as part of this web is an important aspect of relational thinking, which sociology is situated to encourage. Specifically, sociology can benefit university students in three ways: by assessing social and ecological problems, by developing sensitivity to diversity, and by bringing unrecognized connections into focus. In turn, these can aid in furthering the cause of social and ecological justice.

Most introductory sociology courses provide an overview of the discipline and its contemporary trends, and may also include historical background, a lineage of founding figures and a survey of theories and methods. These elements are important for students entering a major in sociology. However, an introductory course often becomes part of a liberal arts core curriculum not necessarily intended for sociology majors, and which may be the only sociology course that students take in their undergraduate education. A course for majors is less useful for students who do not proceed into the major. This suggests an opportunity to teach sociology in a way that while still providing a foundation for the discipline also illustrates relational thinking. Teaching sociology relationally utilizes materials for an introductory course, but the focus is less on the discipline and more about the relational worldview of sociology.

Sociology deploys a form of relational thinking to explain the interconnectedness of individuals and groups, which has implications for how we comprehend the social world and our role within it. Beyond transmitting the discipline's accumulated findings, teaching sociology relationally can help students to develop problem-posing and problem-solving skills. To understand problems, sociology moves beyond direct causality in order to recognize that problems may result from choices and decisions in different times and places. This is not to ascribe or deflect blame; it is to better understand how today's solutions might create future problems. Although specialists work on specific topics, and sociology is often taught as such, its general tendency to focus on relationships presents opportunities to introduce the work of disciplinary specialists to those who can benefit from their insights but who do not enter the major.

A key feature of the sociological viewpoint involves distinguishing between common sense and systematic inquiry. Sociology addresses questions about the relationship between individuals and groups, how they relate to and impact one another in a reciprocal way, and how the groups of which we are a part inform the way we understand ourselves and others. Sociology is therefore well-positioned for helping us to face shared problems by considering the consequences of choices made beyond the immediate setting within which a problem is experienced. While both address problems, sociology differs from common sense in ways that can help students to understand how the discipline works (Bauman and May 2001). For example, when discussing problems among acquaintances, casual speech suffices. To gain a clearer picture of the world and our shared problems a more precise way of speaking is necessary, and so as students learn the sociological way of thinking they are also becoming familiar with the concepts and vocabulary of the discipline.

A distinction between sociology and common sense is in the range of experiences drawn upon to understand problems. We may initially understand problems according to individual experiences and assumptions. Sociology expands our thinking by systematically referring to a broader range of experiences that help to pose problems in a way that can inform sensible solutions. The key to this is the way that sociology understands causality. Common sense thinking often limits causality to the immediate context within which a problem is experienced, filtered through personal preferences; sociology moves beyond this by considering the possibility of multiple causalities.

Sociology provides a method of bridging the gap between itself and common sense through "defamiliarization" (Bauman and May 2001, pp. 10-11). This involves a conscious effort to comprehend the world and ourselves beyond familiar cultural assumptions and social settings in order to "open up new and previously unsuspected possibilities of living one's life with others with more self-awareness" and to help "render us more sensitive and tolerant of diversity." Such a sensitivity can help address problems by avoiding unintended outcomes of short-sighted problem-solving, such as scapegoating and social pathologies that are destructive of relationships and justice.

As an illustration of defamiliarization relevant to fostering eco-justice, we can take the problem of environmental destruction. Sociology helps us to consider the implications of taking for granted an unsustainable way of life and the impact that this way of life is having upon ourselves and the eco-systems we share. In discussing environmental destruction and its connection to a now familiar lifestyle, the question of freedom may arise as a way to defend that lifestyle. Democratic societies often take freedom for granted, particularly where the autonomous individual is the primary locus of being. Sociological thinking problematizes the notion of the autonomous individual by demonstrating that individuals are connected in myriad ways, and that over time and space the resulting interdependencies complicate the idea of individual freedom.

To further illustrate, industrialized consumer societies take great lengths to protect private property. However, claiming property as one's private domain also deprives others from accessing it; by saying "this is mine," one is also saying "this is not yours."

Seen in this way, private ownership is a social act because it hinders access by others, which helps students to understand that valorizing individual freedom ignores its impact on the world and others. Teaching sociology in this way can therefore play a role in helping students to understand the environment not in terms of private property that we can freely do with as we please, but rather as a commons upon which we all depend, past, present and future, and the care for—and destruction of—which we are all responsible. Sociology can help to develop a worldview of "we" rather than "me," which in turn may help roll back the destructive selfishness of rampant individualism and interject a sense of justice into shared cultural values and familiar social mores.

Referring to language as the foundation of developing relational thinking, freedom can be understood as a noun or verb. When nouns and verbs are conflated, confusion and conflict can arise, in this case as a result of misunderstanding the difference between the idea of free will and acting freely. Relational thinking provides a way to clarify this linguistic conundrum by demonstrating that to act freely is not solely a choice of the autonomous individual; acting freely depends upon the network of relationships within which an individual is nested. By highlighting the tension between the idea of free will and acting freely, relational thinking is not dismissing freedom; on the contrary, relational thinking can help students to reflect on the responsibility to balance freedom with social and ecological justice.

To take another example, consider our dependency on the automobile. When first developed, the automobile was seen as a neutral technology to ease transportation and enable individual freedom and mobility. Thinking about the automobile relationally provides a way to understand the problems faced by societies that have familiarized its use. Air pollution, noisy traffic and the long term costs of resources such as oil and rubber have made the automobile a central influence upon—even a liability for—the societies involved with this dependency. While Japanese automobile manufacturers have taken steps toward easing the concerns about air pollution and oil dependency through marketing hybrid vehicles, the burden to solve the problems brought about by automobiles has been shouldered by today's manufacturers and policy makers. Thinking about a problem relationally adds another dimension: whatever the problem, it is not entirely the result and responsibility of individuals living today. The case of the automobile is related to the invention, marketing and adoption of a new technology in the past. The problems brought about by automobiles have proceeded from those living in the past who made short-term decisions on its development and use. Similarly, those who presently manufacture, market and drive automobiles are implicated in the problem by focusing on individual autonomy in transportation or generating greater profits in business. While we cannot change the past, we ought to be able to come to terms with its influence on the present. This, in turn, may help students to realize that solving today's problems, if not done carefully and collectively, might bring about problems for future generations. By considering the past, present and future when adopting a new technology, relational thinking can assist in bringing into focus how problem-posing and problem-solving skills are dynamic and interrelated social processes. Sociology, in effect, teaches that we are collective decision makers.

Seeing society and individuals as relationally intertwined overlaps with a trend among university administrators, employers and policy makers who are calling for students entering the workforce to possess problem-posing and problem-solving skills. Often referred to as "transferrable skills," the dual focus on problem-posing and problem-solving replaces a previous singular focus on problem-solving (Progler 2010). While solving problems remains important, a weakness emerged in this singular approach because posing problems was left to authority figures. Encouraging relational thinking in university education dovetails with an emerging recognition that workers, as well as citizens, can engage with a society, an institution or a company in ways unforeseen by managers. Beyond furthering the disciplinary discourse, teaching sociology as relational thinking contributes to providing students with a work-related skill that can also help them to practice democratic decision making, a key element of social and ecological justice.

Cultural Anthropology: Expanding Awareness of Local and Global Connections

Beyond the insights that sociology provides into urban technologically complex lifestyles, there are other ways that the social sciences and humanities can help students to develop relational thinking. While sociology is closer to home, cultural anthropology often focuses on studying distant "others." A typical introduction to cultural anthropology includes a history of the discipline, contributions of influential figures and an overview of theories and methods; as with sociology, these are useful for majors. However, when offered as part of a liberal arts core curriculum cultural anthropology helps students to think relationally by focussing on the affinity between a familiar self and an unfamiliar other and by highlighting ways that local and global are intertwined. Although this section will take us outside the current comfort zone of urban industrialized society, it is equally important because it illustrates a second path by which the social sciences and humanities can be approached as relational thinking, and offers another way to reflect on the project of social and ecological justice.

Although cultural anthropology is a diverse discipline that evolved over the past century, it originated with the study of tribal peoples. Many of the insights, theories and methods of cultural anthropology owe a debt to these peoples, because without tribal peoples the discipline would not be what it is today. At the same time, even though cultural anthropology has moved beyond studying tribal peoples into other specializations, the early and ongoing research on tribal peoples can help to illustrate relational thinking. In what follows, I will survey how tribal peoples are living with a relational worldview and suggest ways that students might come to see themselves as related to—rather than separate from—the lives of these distant others.

Before proceeding, it is important to outline three points about studying tribal peoples. First, they choose to live the way they do. Second, an absence of complex technology does not indicate a lesser intelligence. And third, industrialized society has severely impacted tribal cultures. As Maybury-Lewis (1992, p. xiv) suggests, by coming to terms with these three factors cultural anthropology paves the way to encourage a "mutuality of thinking." The central message of cultural anthropology is understanding without prejudgment, and so inquiry into other cultures begins by looking at ourselves. Modern university curricula bracket off this interrelated quest for knowledge of the self through the other by emphasizing attainment of skills for the industrialized workforce. The knowledge promoted by the education we value is a window onto ourselves and our limited view of the world, which has become less a place of wonder and reduced instead to a quest for certainty and personal gain. While sociology may proceed from and at times conform to this limited worldview, cultural anthropology purposefully diverges from it through systematic, comparative, reflective and thematic inquiry. It asks what we can learn from others and how their cultures embody wisdoms marginalized by our adherence to short-sighted scientific and material advancement.

Through teaching introductory cultural anthropology courses in Japan, and elsewhere, I have found that an obstacle to appreciating the wisdom of tribal peoples is embedded in the term "hardship." When students are exposed to tribal cultures, they tend to judge those cultures from their own taken-for-granted assumptions and experiences of living in a technologically complex and consumer oriented society by dwelling on what they see as hardship or the inconveniences, limitations and difficulties of how tribal peoples choose to live. Focusing only on these preconceived notions overlooks opportunities to learn from others. To frame tribal cultures in terms of what they lack according to our understanding of needs is potentially damaging to our relationship with others, and thus to ourselves, and does little to advance the cause of social and ecological justice. The ethnocentric notion of hardship, then, has to be faced and eventually left behind for a meaningful and sustained inquiry into the lifestyles of tribal peoples to proceed. Furthermore, ethnocentrism tends to polarize our understanding of, and relationship to, tribal peoples between romanticizing and missionizing. We romanticize tribal peoples when we lament our own lost innocence, or yearn for the presumably pure and simple life they seem to represent; we missionize them when we act upon the well-intentioned impulse to bring them out of hardship and into our industrialized lifestyle. Both views are destructive of mutual understanding and justice, because they are considering only ourselves. A way out of the dichotomy between romanticizing and missionizing is to listen. Cultural anthropology developed as a discipline dedicated to systematically and carefully listening to others as a way to learn about the self in relation to the other.

Once we move beyond the obstacles to learning about tribal peoples, we can take steps toward seeing how their cultures are intimately intertwined with their ecosystems and recognize how their daily lives are enabled through relational thinking. Perhaps the clearest illustration of how tribal peoples see and live in the world relationally is through Bateson's (1972) idea of "ecology of mind," which has parallels in several tribal cultures. The nomadic Gabra of Kenya, for example, live by way of *finn*, which can be translated as "ecology of mind" in the general sense that consciousness and environment are interrelated, but which also specifically refers to recognizing abundance, fertility and the cycle of life. A similar sensibility is found among the Australian Aborigines, who express it with the pithy aphorism "nothing is nothing," while in North America Chief Seattle famously proclaimed, "Whatever befalls the earth befalls the sons of the earth. Man did not weave the web of life, he is merely a strand in it. Whatever he does to the web, he does to himself" (in Maybury-Lewis 1992, p. 62). While our scientists can and do warn us about what we are doing to ourselves and our environment, we may not heed their advice, especially when it becomes expensive or inconvenient.

One way to deepen ecological and human connections is to listen to those who we perceive to be deprived, which beside being an exercise in humility can open up possibilities for learning how to think relationally. Often considered marginal or anachronistic compared to urban dwellers, those who follow nomadic lifestyles, such as the Gabra, are bound on a daily basis to the sobering realization that there are limits within which they must live. Their keen recognition of the limits of ecosystems to sustain life is the key to their survival and sense of eco-justice, and through realizing natural limits they have developed an intimate yet practical relationship with their environment. The practice of hauling water by the bucket from deep desert wells, for instance, may appear to outsiders as a hardship that could be alleviated with mechanical pumps. But the assumption that the Gabra are deprived of labor saving devices occludes a subtle cultural wisdom which recognizes that to adopt such devices would be culturally destructive and alter their relationship with water. Hauling water manually, in addition to inculcating a respect for every drop, turns the well into a place of community. While we have adapted nature to suit our needs, nomads have adapted themselves to nature. A careful adaptation to limits has led the Gabra and other tribal peoples, who we may perceive as lacking technology and living in hardship, to develop sustainable cultures and vibrant communities that stem from a close relationship with their ecosystems.

Reflecting on relational thinking among tribal peoples also highlights the importance of nurturing connections with nature beyond what can be developed through common sense and personal experience. Just as we are guided by scientists, tribal peoples are guided by shamans. Shamanistic ways of knowing are only recently being understood through the work of ethnobotanists, who are coming to terms with the tribal wisdom of seeing ecosystems beyond their value as resources; they are also sources of knowledge that are nurtured through careful and patient observation. Tribal wisdom and modern botany both share a systematic way of seeing nature in an ecological and mutually interdependent framework. This view will help us to better understand "primitive" hunter-gathers who do not see themselves as having dominion over nature, but who have learned to live in a reciprocal relationship with nature. Sometimes, this view is formalized as a pact with nature, as found among the Xerente of Brazil, who tell stories that exemplify reciprocity between hunter and hunted, or the Makuna of Colombia, who believe that reality is singular and that animals, plants and other beings have their own communities and share a spiritual connection. Industrialized societies ought to learn from such cultures about how to develop a moral commitment to seeing survival as intimately related to nature and our fellow creatures. Beyond ecological concerns, this can inform psychological well-being. As Chief Seattle once said after observing the mass slaughter of the buffalo by Europeans: "If all the beasts were gone, man would die from a great loneliness of spirit. For whatever happens to the beasts soon happens to man. All things are connected" (in Maybury-Lewis 1992, p. 59).

One might feel that examples such as these are impractical, owing to concerns with teaching job skills. It may indeed be practical for universities to provide students with the skills to succeed in a market driven economic system. This overlooks the impact that competitive social relations can have on well-being and belongingness, and that a market dominated worldview is preventing us from appreciating the wisdom of others.

Relational thinking can help illustrate that not all economic activities are governed by efficiency. Seen from a market-driven perspective, the perilous seafaring journeys of the Trobriand Islanders to exchange items that appear to have no market value seems to lack logic. However, as several anthropologists have noted, this assumption ignores the practice as a system of reciprocal gift giving that is intended to build relationships (Malinowski 1922, Mauss 1954). Among the Weyewa of Indonesia, rather than using machines to cut, move and set an enormous memorial stone, a community elder will muster his network of relationships to complete the task by hand and manual labor. Such a man, while seeming to be poor if viewed in material terms, is rich in relationships. As one elder, Lende Mbatu, put it, "I am not a rich man according to most human reckonings but I am rich in ability and I am rich in knowledge, I'm rich in favors and I'm rich in cooperation with others" (in Maybury-Lewis 1992, p. 72). Economic activities such as these bring people together in reciprocal exchanges in which it is desirable to be indebted to one another because being in debt perpetuates relationships. The Gabra illustrate this sense of reciprocal debt. While camels are sources of food, clothing and transportation, their main purpose is for giving and lending. Such a culture is disrupted if someone acts in self-interest. A Gabra camel herder put it this way, "Even the milk from our own animals does not belong to us. We must give to those who need it, for a poor man shames us all" (in Maybury-Lewis 1992, p. 85). Such seemingly impractical views indicate a moral economy prioritizing relationships.

University students are often preoccupied with finding themselves, with questions of identity. Cultural anthropology suggests that most societies, large or small, fit members into an image of personhood. Among tribal societies, identity is based on creating social and cosmic relationships rather than demarcating the autonomous individual. The Xerente, for example, do not name their children until several years after birth. Initially, Xerente children are referred to with generic terms such as daughter, son, niece or nephew. Naming is a very serious undertaking because it is "a way of making an individual into a social being, of linking him with society, just as society is linked to the cosmos" (Maybury-Lewis 1992, p. 121). Forging these social and cosmic links takes time and due consideration. Socialization—and thus naming—in tribal societies evolves through interaction. While we do this, too, creating a sense of self is often individually cobbled together amidst the destabilizing pressures of market and media.

Cultural anthropology can also illustrate relational thinking by providing insights to perceptions of reality among tribal peoples. In the Dreamtime of the Australian Aborigines, reality is whole in ways that our sciences may not appreciate. The Dreamtime is the domain from which all has emanated and to which all remains connected; it is the world of the ancestors. Parallel to the seen world in which we all live, the Dreamtime assumes a transient interrelatedness with the unseen. To maintain connections and stability it is necessary to perform rituals, such as dancing along "songlines" that form a cosmic map of the terrain. While they appear on the surface to be community activities, with rituals inculcating bonds among members, they also maintain bonds with the Dreamtime and with an ecosystem. Aboriginal peoples understand past, present and future, the seen and unseen, as intimately interrelated.

Teaching cultural anthropology as suggested in this brief survey can help to make distant others intelligible. In so doing, anthropologists have helped us to see tribal peoples not only in terms of their differences from ourselves, but also through their similarities to ourselves. Some of these similarities include how identities are socially constructed or in the way gift-giving builds relationships. But perhaps most intriguing of all is the ironic similarity between the ancient Dreamtime and modern quantum physics in that both apprehend reality as needing our participation to make it manifest. University students may not fully comprehend the mysteries of quantum physics or the intricacies of the Dreamtime, nor may they wish to dwell in "hardship" among tribal peoples, but teaching relational thinking through cultural anthropology might help our students to develop an appreciation of the relational nature of the world and gain some respect for those peoples who choose to live in ways different from, yet akin to, ours.

Comparative Religion: Cultivating Intrapersonal and Interpersonal Relationships

As we have seen with sociology and cultural anthropology, relational thinking can be illustrated by emphasizing selected aspects of introductory course materials that bring to the forefront socio-cultural connections. While sociology helps with integrating problem-posing and problem-solving skills and cultural anthropology assists in expanding awareness of local and global connections, comparative religion offers insights to cultivate two kinds of relationships. Firstly, comparative religion pays attention to the intrapersonal dimension, specifically in how we understand, from a metaphysical standpoint, ourselves and our place in the cosmos. Secondly, comparative religion considers possibilities for developing interpersonal relationships through studying various ethical precepts on how to live life with others.

Before proceeding to the more familiar historical religions, we will remain with tribal peoples and explore the metaphysical dimension. A distinguishing feature of tribal cultures is living in an oral-based, rather than text-based, world. Orality bolsters community because of its reliance on memory and face-to-face communication. University students tend to rely less on memory, and our educational system values information and texts abstracted from speech and shorn of cultural context. Emphasizing utilitarian skills in university curricula further marginalizes memory, which is erroneously thought to be replaceable by databases, thus making the autonomous individual the primary locus of understanding. For tribal peoples knowledge resides in collective memory; to seek knowledge it is necessary to maintain relationships. While memory may appear to have limitations, and libraries and databases might seem more expansive and reliable, with a proliferation of abstract information we are losing the ability to determine what is important. The natural limits of human memory encourage tribal peoples to develop a clearer sense of what is important and what is superfluous. Setting aside printed texts and abstract data as the basis of knowledge and learning can also open up the senses to the sacred, and it is with this connection between orality and the sacred that we will begin our discussion of tribal religions.

To grasp the religious outlook of tribal peoples, students can be encouraged to ponder the difference between place and space. As the dominant trope in our world, space is quantitative and movable. While certain spaces take on value as real estate, space itself is mobile. A square meter remains as such wherever measured, even as its monetary value fluctuates. This abstract sense of space differs from the tribal understanding of place, which has specific ancestral, cultural and spiritual qualities that are handed down through face-to-face oral communication and which bind people to a locality. Tribal peoples are intimately rooted to the places in which they live and these places are sacred. A particular mountain, forest, lake or valley is not interchangeable with others; each has specific experiences, memories and mysteries attached to it. Smith (1994) relates the story of Haudenosaunee chief Oren Lyons, who upon returning from his university studies away from the tribal lands was reminded of this attachment by an uncle as they sat in a canoe in the middle of a lake. His uncle pointed out that Oren *is* this place, that he *is* the lake, the mountains and the trees, that he is *embedded* in this specific place. To collapse such a meaningful relationship with place into measured space governed by unstable markets and abstract texts is to unweave the fabric of tribal spirituality because it disrespects their sense of ecological relationships, which brings about misunderstanding and, ultimately, injustice.

As with place, relational thinking is evident in tribal notions of time. Industrialized societies tend to mark time incrementally and linearly, whereas tribal peoples live in an atemporal world that encompasses the past, present and future, and thus signifies more concern with causality than chronology. Paradoxical to our chronological sensibility, tribal peoples do not see the past as distant because it is actually closer to the spiritual origins of their world. Seeing time as an atemporal bundle requires vigilance to keep the world in order, as we saw above with the Australian Aborigines. Enacted through ritual, maintaining a relationship with their origins has a metaphysical meaning and is an essential feature of how tribal peoples experience the sacred. Since in many tribal cosmologies animals originated before humans, they are closer to the sacred and therefore can provide spiritual guidance as totems. Animal, mineral and vegetable are sacred and interconnected in a vision of time and place that differs significantly from our way of seeing ourselves in the world. Tribal religions therefore have a feeling for immanence that becomes manifest through relying on what Smith (1994, p. 241) refers to as a "symbolist mentality" that "sees the things of the world as transparent to their divine source; whether that source is defined or not." This contrasts with our materialist mentality, which "recognizes no ontological connection between material things and their metaphysical, spiritual roots." For tribal peoples the sacred, and thus their religious sensibility, is ecological and experiential.

To complement the immanence of tribal religions, the historical religions are defined by transcendence, the pointing to a unified, perfect and infinite world that is beyond ourselves and which can never be fully experienced. As another key attribute of the religious sensibility, the wisdom here is in the pointing itself, rather than in the goal. Infinite love, perfect grace, supreme justice, striving for these, even if falling short, makes one a better person. To encourage this betterment, the religions develop ways to narrate, such as through parables, or depict, such as through art, these essentially unattainable states. As with religion, our sciences also point to worlds beyond direct experience, such as the quantum universe or distant galaxies, and have developed ways to render these worlds knowable. The difference is that religions derive ethical precepts from pointing to these unseen worlds. Ultimately, the view of comparative religion toward transcendence reveals a way of knowing based on interrelationships between what appears on the surface to be a fragmented imperfect reality and a perfect unified reality residing beyond direct perception. Our inability to prove that this transcendent unity exists might be seen as an obstacle to understanding, but that is only because we tend to apply the preconditions of empiricism to something that is essentially mysterious.

This is not to denigrate science, nor to champion religion. The scientific method, including that which is deployed by the social sciences, has led to admirable and useful observations on the nature of things. However, this is done by isolating and examining fragments for answers. At times, a bigger picture is inferred from these disintegrated fragments, but this is not a requirement for doing good science. Religion, on the other hand, begins with the view that "things are more integrated than we normally suppose" (Smith 1994, p. 248). Not constrained by the scientific method and the parameters of an empiricism that restricts itself only to that which can be measured, religions tend to work in the opposite direction by beginning with an integrated view and then looking at the details in light of it. Focusing on this vision of the integrated unity of existence—whether immanent or transcendent—has implications for pondering social and ecological justice because it suggests how comparative religion can help in learning about intrapersonal and interpersonal relationships.

The Hindu tradition, for example, actualizes the intrapersonal dimension of life by urging a move beyond self-centeredness and embarking upon the path of renunciation. One way this can be achieved is through karmic yoga, which includes a focus on selfless community service. Distinct from the contemporary practice of yoga in the West as bodily regimen, in Hinduism yoga is a spiritual technology aimed at joining the finite self and the infinite within. As one of four varieties of yoga, karmic yoga is a form of relational thinking in practice that proceeds from an awareness that human acts are connected beyond the physical and temporal realms in which they occur.

The emphasis in Hinduism on transcending the self to achieve an integrated oneness provides an indication of relational thinking among people whose lives are informed by traditional religion while living in modern industrialized societies. However, we cannot extend this recognition as the defining feature of Indian society. Rather, it is a possibility that may be selectively cultivated by members of a society. Similarly, it would be facile to say that Zen Buddhism somehow represents Japanese society, but we can suggest that Zen provides the option to labor toward an existential oneness. Through its focus on how language and reason cloud enlightenment, Zen encourages its adepts to negate the rational truth-seeking mechanisms of the mind and enter a direct communing with the universe. This differs from philosophy, which relies on argumentation, logic and rational language; Zen attempts to clear these as blockages to getting in touch with the ultimate reality. Students do not need to undertake the rigors of yoga or apprentice themselves to a Zen master to appreciate how religions develop intrapersonal relationships, as we can teach comparative religion in a way that suggests a relational view of an intrapersonal self.

An illustration of relational thinking as it pertains to interpersonal relationships can be found in Confucianism, a defining feature of Chinese, Japanese, and other East Asia societies. Since Confucianism is included in textbooks on comparative religion, we may put aside the academic debate on whether it is a religion, philosophy or ethical system since we are interested here in what it can demonstrate about relational thinking. The ancient teachings of Confucius have become so ingrained today that they are seen as natural features of many East Asian societies. As a taken-for-granted way of being, Confucianism has gone beyond explicit faith and become implicit culture. Most people living in the Confucian societies are not concerned with the origin of this tradition since its values have for centuries formed the normative basis for interpersonal relationships. At the center of the Confucian sense of values lies empathy, putting oneself in the place of others, and seeing the individual as nested within concentric circles of interactive empathy. This can partly explain why East Asian social relations are bound up with the group, a type of collectivism in contrast to Western individualism (Nesbitt 2003).

We could extend this discussion into the more familiar monotheistic religions and find similar aspects of how transcendence brings existential unity and ethical precepts bring social harmony and justice. One would not have to look far to find, for example, a connection between these traditions and our sense of justice today. The monotheistic religions emphasize what Cornel West has called the "prophetic voice," a well-known example being Dr. Martin Luther King struggling for racial equality and social justice by evoking the story of Moses, which is incidentally common to all three monotheistic faiths. What matters for our discussion here is that, as with the other religions, through sensitive inquiry relational patterns emerge. The challenge, which is beyond the scope of this chapter but which I have addressed elsewhere (Progler 2014), is to proceed in a non-sectarian way that takes the religions seriously and on their own terms and which highlights the similarities among them as well as the differences between them.

Finally, comparative religion illustrates a way to think about the relationship between faiths. The religions of East Asia have tended to be configured as complementary in the sense that they are related to, rather than being separated from, one another. In pre-modern China and other parts of Asia influenced by China, as well as to a great extent today, Confucianism guides interpersonal relations, Taoism informs health and hygiene, and Buddhism cares for death and remembrance. An old Chinese saying is apt here: Every black-haired child of Han wears a Confucian hat, a Taoist robe and Buddhist sandals. In Japan, this is encapsulated in a similar way: Japanese are born Shinto, marry Christian and die Buddhist. Relations between religions in East Asia are therefore characterized by "multiple affiliation," in that each religion has something to offer and that they can relate to one another as partners because each speaks to a different facet of the human condition. This both/and view differs from the West Asian way of understanding relations between religions, which is based on "exclusive affiliation" and which suggests an either/or configuration. While religious affiliations in the West are diverse today, we have tended to adopt the West Asian configuration in both religious and secular outlooks, the latter including atheism, so perhaps the resulting conflicts can be tempered by considering the East Asian configuration. Beyond these particularities, the most important observation from comparative religion is its insistence on non-judgmental inclusiveness, that we make room for differing perspectives and that these can reside alongside one another as part of an implied whole. It is akin to climbing a mountain; while the peak is one there are different paths to reach it. This holds a lesson for us all on the importance of relational thinking for putting into practice a sense of equality and social justice, which we ought to be passing on to our students in the classroom.

————

With a realization that the liberal arts provide opportunities for students to become acquainted with general academic skills and knowledge about themselves and the world in which they live, relational thinking can become a unifying strand in higher education. This is particularly the case in complex culturally diverse international settings within which students and faculty from different backgrounds come together. Sociology, cultural anthropology and comparative religion can be approached as forms of relational thinking with each making a distinctive contribution. The more metaphysically oriented macro-view of comparative religion complements the ethnographic micro-view of anthropology, and both are complemented by the sociological perspective. In addition to learning how the disciplines operate, students may gain an understanding of relational thinking as a factor in ecological and social justice. By re-configuring the available teaching materials, relational thinking can be highlighted to provide an interdisciplinary viewpoint in the liberal arts. This is not to undermine the integrity of the academic disciplines, much less to advocate their replacement. Rather, like all vibrant intellectual traditions, and as many have done before and continue to do so, the academic disciplines of the social sciences and humanities can evolve and find new meaning within the academy.

❖ ❖ ❖

Joseph Progler teaches Culture, Society and Media at Ritsumeikan Asia Pacific University in Japan. He has taught International Studies at Zayed University in Dubai and Social Studies Education at Brooklyn College in the City University of New York, and is a co-creator of Multiversity, based in Penang and Goa. His writing is available at https://ritsumeikan-apu.academia.edu/JProgler

References

Alen, O. "Rhythm as Duration of Sounds in Tumba Francesca." *Ethnomusicology*, Vol. 39, No. 1, Winter 1995, pp. 55-71.

Bateson, G. *Steps to an Ecology of Mind*. Chicago: University of Chicago Press, 1972.

Bauman, Z. and May T. T*hinking Sociologically*, 2nd edition. Oxford: Blackwell Publishing, 2001.

Berliner, P. *Thinking in Jazz: The Infinite Art of Improvisation*. Chicago: University of Chicago Press, 1994.

Burnside, J. and Baker, N. *Relational Justice: Repairing the Breach*. Winchester: Waterside Press, 2004.

Carpenter, T. P., Levi, L., Franke, M. L. and Zeringue, J. K. "Algebra in Elementary School: Developing Relational Thinking." *Zentralblatt für Didaktik der Mathematik*, Vol. 37, No. 1, February 2005, pp. 53-59.

Date, C. J. *SQL and Relational Theory*, 2nd Edition. Cambridge: O'Reilly Media, 2011.

Debaise, D. "What is Relational Thinking?" *Inflexions*, Vol. 5, March 2012, pp. 1-11.

Jones, M. "Phase Space: Geography, Relational Thinking, and Beyond." *Progress in Human Geography*, Vol. 33, No. 4, August 2009, pp. 487-506.

Keil, C., "The Theory of Participatory Discrepancies: A Progress Report." *Ethnomusicology*, Vol. 39, No. 1, Winter 1995, pp. 1-19.

Kineman, J. "Relational Science: A Synthesis." *Axiomathes*, Vol. 21, No. 3, 2011, pp. 393-437.

Malinowski, B. *Argonauts of the Western Pacific*. New York: Routledge Classics, 1922.

Mauss, M. *The Gift: The Form and Reason for Exchange in Archaic Societies*. New York: W. W. Norton and Company, Inc., 1954.

Maybury-Lewis, D. *Millennium: Tribal Wisdom and the Modern World*. New York: Penguin Books, 1992.

Nesbitt, R. E. *The Geography of Thought: How Asians and Westerners Think Differently… and Why*. New York: Free Press, 2003.

Papadatou, D. *In the Face of Death: Professionals Who Care for the Dying and the Bereaved*. New York: Springer Publishing Company, 2009.

Progler, J. "Searching for Swing: Participatory Discrepancies in the Jazz Rhythm Section." *Ethnomusicology*, Vol. 39, No. 1, Winter 1995, pp. 21-54.

Progler, J. "Curriculum Reform in the Corporate University: From the Disciplines to Transferrable Skills." *Social Systems Studies*, Vol. 21, September 2010, pp. 95-113.

Progler, J. "Internationalization and Cultural Diversity in Higher Education: Teaching for Mutual Understanding." *Ritsumeikan International Affairs*, Vol. 12, March 2014, pp. 61-82.

Rosen, R. *Life Itself: A Comprehensive Inquiry into the Nature, Origin and Fabrication of Life*. New York: Columbia University Press, 1991.

Schulter, M. and Lee, D. *The R Factor*. London: Hodder and Stoughton, 1993.

Smirnova, A., Zorina, Z., Obozova, T. and Wasserman, E. "Crows Spontaneously Exhibit Analogical Reasoning." *Current Biology*, Vol. 25, No. 2, 19 January 2015, pp. 256-260.

Smith, H. *The Illustrated World's Religions: A Guide to Our Wisdom Traditions*. New York: HarperCollins Publishers, 1994.

Stephens, M. "Describing and Exploring the Power of Relational Thinking." In P. Grootenboer, R. Zevenbergen and M. Chinnappan (Eds.), *Identities, Cultures and Learning Spaces, Proceedings of the 29th Annual Conference of the Mathematics Education Research Group of Australasia*. Canberra: MERGA, 2006, pp. 479-486.

Two Faces Of Eco-Justice In Chinese Society:
De-Capitalizing Schooling Reform For A Sustainable Future

Chun-Ping Wang

1. Introduction

It is apparent that ecological problems are closely related to the degree of modern civilization development. As the public's daily lives become increasingly dependent on digital technology, as well as on organic solidarity maintained by dense labor division, it is not likely that people give up the convenient lives served by different service trade. In order to keep such comfortable lives and to maintain their consuming abilities, people can only work hard. It is exactly the economic based consuming culture that makes us on one hand pursue efficiency, progress and excellence, the values that capitalism advocates. And on the other hand, it makes us depreciate traditional values. This attitude is the root of all ecological problems.

If we gain prosperity at the moment by paying a price of a sustainable future in this Faustian deal, it is understandable for human are inclined to pursue happiness and avoid pains. When most people tend to value their personal enjoyment more than environmental conservation, the natural environment becomes an object of ruthless exploitation. In other words, when economic growth is interconnected with each individual's real life, and people care about "how to keep a comfortable life" or "how to transfer environmental problems to others", it brings about capitalization progress. The chase of "self-interest" demonstrates the thinking of efficientizing; nevertheless, it causes conditions of eco-injustice. Weighing upon the basic rule of capitalism, that is the premise of limited resources and infinite desire, when each individual wants to maintain certain lifestyle, the simplest solution is to reduce production cost to make commodities affordable to everyone. Nevertheless, how do we whittle down production cost? Surely the simplest way is to move factories to places with cheap labor, for instance, Coca Cola moved the factories to India, and transferred jobs of planting coffee to South America. Nike sports shoes even exported the whole production line to South Asia. Some advanced countries also import cheap raw materials from economically backward countries, such as importing corns to produce biomass alcohol, ostensibly to solve problems caused by global warming. Some may argue that moving factories to the Third World is a win-win relationship; this is tantamount to let the local poor being free from starving and the development of biomass energy carry out the duties as global citizens. From the "double bind" point of view (Bowers, 2011), there will be more than one facts—moving the production line to the Third World will certainly transfer the environmental problems to the local areas, and the development of bio-energy may increase the costs of the poor in buying foods. Problems arising from this situation are not merely a social justice issue, but a more "ecological justice" issue.

The term "ecological justice" was originated from the book "Environmental Justice" written by an American scholar, Peter S. Wenz, in 1988. He advocated that one should integrate the theory of justice with ecology studies. However, ecological justice is different from "environmental protection" and it is distinguished from environmental justice (Wenz, 1988). In general, environmental protection is the narrowest concept; it only refers to the minimal environmental damage caused by human. In addition, some scholars used the terms "environmental justice" and ecological justice interchangeably, extending the common concerns of environmental justice or ecological justice from human-centered to life-centered or ecology-centered. However, some scholars maintain that the two should be distinguished. They believe that ecological justice belongs to the umbrella of the theory of environmental justice, and emphasize that the need to apply the concept of justice not only occurs among people (narrow environmental justice) but should be expanded to between man and nature (ecological justice) (Hwang & Hwang, 2007). Thus, ecological justice no longer regarded human as the master of all things but an entity share the same ecological space with other lives, and underlining the value of sustainability and symbiosis.

In the United States, the development of environmental protection and ecological awareness, for example, ecological justice is seen as a part of social justice. Being influenced by human right activists, the sites of waste treatment facilities in earlier days in the States mostly located in low-come or people-of-color communities. Gradually, the issue of environmental injustice in the States gained more attention by different parties and was made into a political issue. The phenomenon of locating waste treatment facilities which caused environmental problems in the minority communities has been criticized as "environmental racism" (Hwang & Hwang, 2007). With the expansion of global capitalism, the ecological crisis and environmental racism prejudice present in the earlier American society, now continue to appear as ecological injustice problems in many countries. In his book, "World-systems Analysis: An introduction", Immanuel Wallerstein pointed out that the modern world system is a social entity of labor division, and this division of labor is not only the division of functions but geographical segments. To put it simple, the division of labor within our modern world system is unequal; those who occupied the core dominant areas are often economically developed countries. These core countries, due to unequal labor division, use cheap raw materials and labor of outlying and semi-outlying countries to produce high value-added products and thus monopolize the world market chronically. While joining the global economic system may help the development of backward countries, outlying and semi-outlying countries may gradually move toward the center (Wallerstein, 2004), this is not a bad thing for the poor's lives can be improved by joining the world system, if simply considering the aspect of economy. Yet, the problem of ecological injustice surface when judging from a point of view of long-term development of sustainability.

According to Wallerstein's world-systems theory, the earlier "Four Asian dragons", the "BRIC", or "The rise of China," are all examples of gaining rapid economic development by joining the world-economies. However, economic development is always at the expense of ecological resources in return. The aim of this article is to compare and contrast the definition of "ecological justice" in the People's Republic of China and in Taiwan, and to explain the ecological injustice problems, laws made to protect the environment, and reforms in schooling in each of the countries, in hope to show the different efforts that the two Chinese societies made when facing the need of developing a sustainable future in the 21st century.

2. The two faces of eco-justice: sustainability and the construction of ecological civilization

The world-systems theory suggests that there are two shares of the force, the push and pull forces, during the process when low-capitalized countries join in the global economies. In order to improve the economy, a force will naturally emerge in the less developed countries and urges them moving toward the direction of the core countries. Meanwhile, because of the production techniques and ample funding, rich countries will naturally hold tight the less developed countries. In such a push-pull relationship, the rich countries ensure their dominant position and their people maintain a good quality of life. Whilst, by expanding multinational enterprises investment, the less rich countries aim to increase job opportunities and thereby raising the gross domestic product (GDP). Purely from an economic development perspective, it is almost the inevitable trajectory that every country follows after they joined the global economic system. However, from the perspective of environmental ethics or ecological justice, the enjoyment of people living in rich countries is apparently gained through colonizing the environments.

In terms of Taiwan, it has been called the "Asian dragons" together with Singapore, South Korea and Hong Kong. In the past, Taiwan prioritized agriculture and light industry; but from the 1970s to the 1990s, through the original equipment manufacturer (OEM) products, the abbreviation MIT which means "Made in Taiwan" has thus become synonymous with cheap and low quality. In 1998 the Hollywood movie "Armageddon" depicted the extinction of all life on earth caused by meteorite hitting. To save the planet, a team of scientific experts in oil drilling was sent to ride on the space shuttle and carry out the task of blasting in space, blasting tasks carried meteorites. When the space shuttle tried to connect with the space station and the station's equipment was not functioning, a Russian astronauts flapped the dashboard saying: "Russian component, American component...all made in Taiwan!" This line reflected the general impression the European and American have on Taiwan and their memories of lives filled with cheap and poor quality goods from Taiwan. After joining the line of economically developed countries, and also being influenced by liberalism and political democratization, Taiwan paid more attention to sustainable development and eco-civic literacy:

The idea of "Sustainable Development" is to establish a sense of "global citizen" responsibility. It has two main principles: (1) to meet the contemporary needs without compromising the needs of future generations. That is the aim is to promote and create contemporary well-being, without reducing the welfare of future generations as the cost. (2) to make good use of all the natural resources without reducing their environmental basic stock. That is when making use of biological and ecological systems, we need to maintain sufficient amount for regeneration. (Wang, 2008)

It is obvious that one of the goals of Taiwan's "ecological justice" perception is to promote sustainable development and a common share of the right to the environment. Accordingly, how to avoid infringement of others' right of equal development has become the perfect duty of ecological citizens and green enterprises.

Similar to Taiwan, China is also a Chinese ethic dominated society. China was of a closed social system; began to implement "internal reform, opening-up" policy since 1978. Compared with Taiwan's development transition, during these decades, China's GDP kept growing roughly 9% annually. According to the IMF statistics, China's GDP in 2014, US $17.6 trillion, surpassed the United States' GDP (US $ 17.4 trillion) and became the world's largest economy (Wikipedia, 2016). Taiwan's economic miracle was of course at the cost of the damage of the environment and people's health; while "The rise of China" paid similar prices. In a documentary called "Under the dome" (full name: "Chai Jing's review: Under the Dome—Investigating China's Smog"), Chinese journalist Chaijing published in 2015 a survey of the air pollution in China. The film adopted a similar way to present as what is used by Al Gore in "An Inconvenient Truth". When Chaijing was pregnant, her unborn child was diagnosed with cancer and needed surgery as soon as being delivered. Then she found that the smog is at the root of all problems (Chai, 2015).

Chaijing puts forward some statistics of the smog, such as "living in Beijing which is polluted 175 days a year", or "being afraid that her daughter will one day ask, 'What is the blue sky' and 'Why always keep me in the house'" (Chai, 2015). As a mother, she felt "she must understand what haze is, where did it come from, and what to do next". During the investigation, she found out that in January 2013, there were 25 provinces and around 600 million people in China affected by the smog. And the longerst time that an area was polluted by continous haze was as long as 264 days. Further investigation showed that 60% of the haze was caused by burning oil and coal to produce energy for industrial production. In 2013 China's steel industry has burnt 3.6 billion tons of coal, which is more than the sum of other countries in the world. Although the industry has created hundreds of thousands of jobs, it took 3-6 tons of coal and 600 kilograms of water in order to produce a ton of steel, and emitted 1.53 kilograms of sulfur dioxide. Yet only less than two yuan (Renminbi) were earned; at last, the industry needed to rely on the government subsidies to maintain the operation. Due to the excessive urbanization, averagely 80 villages disappear everyday in China. Over-exploitation resulted in excess production of steel and countless construction sites brought more dust. All these conditions made the air pollution problem in China deteriorating.

In the beginning of reform and opening up, the highly discussed issues were "how to allocate resources to achieve social justice", "how to make the cake bigger", or "how to distribute equally". Yet nowadays every morning as they turn on their mobiles, the main concern of the Chinese public is the information about PM 2.5 of the air quality. They only need to look up at the sky and they will know whether to go out wearing masks (Dong & Zhao, 2015). Facing such serious ecological problem, China for the first time proposed the construction of ecological civilization policy guidelines in the Chinese Communist Party 17th Party Congress report in 2007. It stressed the importance of resources conservation and environmental protection policy (Liu, 2009). According to the explanation of Chinese scholars, the so-called "ecological civilization" is an emerging civilization pattern after the original human society has developed agricultural civilization and industrial civilization. Such state of human civilization maintains prosperity through the harmonious development among mankind, nature, and the society (Zhang, 2009). Considering from this policy direction, China's advocate will follow the "harmonious development among mankind, nature, and the society". The possibility of the "harmony" is based on the premise of the resource conservation and ecological protection. In terms of the form, the concept of "ecological civilization" China put forward seems to be close to the "sustainable development" that Taiwan pursuit, yet there are essential differences between the two. What embodied in Taiwan's definition of "ecological justice" are the values of liberalism and universal human rights; while Chinese-style "ecological civilization construction" is based on the Eco-Marxism:

Eco-Marxist believes that only the abolition of the capitalist system and the establishment of a new type of eco-socialism can solve ecological problems. The profit motive of Capitalism leads to the endless pursuit of maximum production and consumption without regard to the rationality of the development and utilization of natural resources. Such an economic rationality will inevitably damage the ecological environment, causing ecological crisis. Eco-Marxist proposed the establishment of an eco-socialist society, which underlines equality, freedom, and justice within the society. It is believed that such society as a whole being harmonious with the nature will achieve a high degree of ecological civilization (Wang, 2013).

Regarding the political system, China still insisted on one-party communist dictatorship, whereas headed for capitalization in economy. China called this capitalist line as "socialism with Chinese characteristics" in attempt to hold fast the political bottom line of the Marxist camp. From another point of view, China's economic development is controlled by political factors, thus it does not belong to the free market capitalization. Some Chinese scholars defined ecological justice as "different countries, regions, and communities owning fair rights and sharing equal obligations" (Zhang, 2009). This idea clearly includes two aspects of ecological justice; one is that all rights are equal and the other is right violations must be compensated. Taiwan adopted a laissez-faire capitalism, and is concerned about the former, which is active promotion of equal rights. On the other hand, China has adopted limited market economy, with emphasis on the latter, which is the overall growth of the national interest. Due to deliberate avoidance of the solidarity driven by ecological justice awareness, as long as the politics red line is not trespassed, economic development in China remains a priority when it conflicts with environmental protection.

3. The practice of environmental regulations and school education of the People's Republic of China

As China is one of the objects among the developing countries that the developed countries would transfer their environmental deficits, according to statistics, China has imported 990,000 tons of waste in 1990 and imports amounted to 2.6 billion US dollars. In 1997, waste imports has soared substantially to 10.78 million tons, and the import amounted to 2.95 million US dollars. The People's Republic of China has set three types of restrictions for foreign capital business model, including "joint ventures", "foreign cooperative enterprises" and "exclusively foreign-owned enterprises". It is not difficult to imagine that the gradual expansion of economy scale will impact on the ecology. Among the total foreign-funded enterprises in 1995 in China, for example, there were 16998 enterprises invested in high polluting production, which was 30% of the three kinds of the enterprises aforementioned (Wei & Wei, 2008). Therefore, after the economic reform and opening up, the economic development of China was fruitful in a short term. Such economic development level had been achieved by the Western societies all the way through a long process of hundreds of years. But China achieved through a highly time-compressing method. Time compression during the process of development has led to excessive expansion of space and to various social problems. Excessive exploitation of resources in particular impacted directly to the balance of ecological environment (Qian, 2014).

Despite the large scale industry of China has only been developed for half a century, the large population, rapid development and the policy mistakes in the past have resulted in environmental and resource problems. Related problems in different aspects are listed as follows (Ministry of Education of People's Republic of China, 2004):

(i) Water and soil loss: the area of water and soil loss in China is 38% of the total land area

(ii) Expansion of desertification area: the desertified land area in China is now 27% of the total land area, and keeps growing an annual rate of 2460 square kilometers

(iii) Large areas of forest have been cut down: the destruction of natural vegetation greatly lowers its ability in confronting wind, holding sand and soil, keeping water, cleaning the air, and conserving biodiversity;

(iv) Three problems (i.e. degradation, desertification, and salinity) of the grassland deteriorating: the problematic grassland area is one third of the total grassland area in the country and increase at a rate of 20,000 square kilometers per year;

(v) Biodiversity has been severely damaged, accelerating the extinction of species: there are 15-20% percent of plant and animal species in China that are threatened, which is higher than the average level of 10-15% in the world;

(vi) Worsened water pollution: suffering serious shortage of freshwater whilst using existing water resources unreasonably;

(vii) Serious air pollution: acid rain pollution area was gradually expanded due to pollutions; acid rain appeared in most of the southern cities which has become one of the three acid rain areas;

(viii) Municipal solid waste (MSW) pollution, which is also named "white pollution", has become more critical, and the situation that the urban areas are surrounded by wastes increased;

(ix) Urban noise pollution grows;

(x) Radioactive contamination and electromagnetic radiation become potential threat to human health.

These environmental problems are related to different dimensions of eco-justice issues. The different forms of ecological injustice put the minorities in the society and other species at risk. According to analysis of academics, ecological injustice in China nowadays may be divided into three types (Zhang, 2009):

The first type is the urban and rural inequality: the pollution control investment in China was mostly used in industrial development and urban areas, and there are still 300 million people in rural China drinking unclean water and 1.5 million acreages of arable contaminated. Every year, 120 million tons of garbage are stacked outdoor, environmental protection measures in rural areas is almost zero. The urbanization in China is apparently at the cost of sacrificing rural areas.

The second type is regional inequality: for decades, China's resource-rich but less developed areas kept offering resources to the developed regions. Nevertheless, the developed regions failed to provide any compensation to the remote areas. In order to solve the worsening environmental problems, China began to limit the development of the western region, but required the poor and polluted areas bearing alone the responsibility of environmental protection, whereas the eastern coastal cities enjoy the fruits of economic development. In recent years, issues of concern include dispatching water of the southern region to the north and ban of forest logging; yet whilst the western region implemented a large scale forest restoration, it was the eastern cities that benefited from it.

The third type is class inequity: The wealthy enjoy better medical care and reduce harms caused by the environmental pollution, whereas, the poor have no capacity to ameliorate physical problems resulted by the environmental pollution

In order to solve environmental problems, China has set the "Environmental Protection Law", "Water Pollution Prevention and Control Law" and other environmental regulations. Although there is not a clear consensus concerning what should be included in these basic laws for environmental protection, a number of specific regulations of obligations of environmental protection are proposed. For instance, the "Environmental Protection Law" Article VI stipulates that all units and individuals shall have the obligation to protect the environment, and have the right to report and accuse any individuals who pollute and do damage to the environment. The "Water Pollution Prevention and Control Law" Article ten also regulates that all units and individuals have a responsibility to protect the water resources and the right to report activities of damaging the water environment. From the provided statutory provisions, however, we realize only the passivity is emphasized in the regulations, that is, environmental obligation within the provisions is only negative obligation (Liu, 2009). A more proactive ecological-based ideal and awareness of ecological citizenship, which respects the right of all species or environmental restoration, are not highlighted.

Confronting the current problems of ecological injustice, China put forward "ecological civilization construction" and hoped to educate the people in China to respect and protect the nature. Environmental problems have to be solved on the basis of reaching the harmony among mankind, nature, and the society, and at the same time we should establish sustainable production and consumption patterns, thereby reaching the interdependence, mutual benefits, coexistence and communion goal of human and the natural environment (Cai, 2015). In the Chinese Communist Party Congress eighteenth meeting, ecological civilization has been expounded more deeply and further emphasized:

> The concept of ecological civilization must be established which includes respect the nature and being in harmony with the nature. Education of ecological civilization should also be propagated to increase the public's awareness of energy-saving, environmental protection and ecology to form rational consumption habits and to create a morale of caring for the environment (Qian, 2014).

In terms of the implementation of environmental education, in 1979 the Chinese Environment Committee has recommended that primary and secondary school education should include environmental education in school curricula. At the same time the committee has also selected some provincial and municipal schools to pilot the curriculum. Since then, the environmental education was included in the basic education system. After several years of pilot, the Chinese National Environmental Protection Administration and the National Education Commission jointly held a national conference in 1985. The National Education Commission resolved that environmental education is statutory curriculum in high school elective courses or extra-curricular activities since 1991 August, and in 1993 it should be extended down to compulsory education. In terms of the teaching of environmental education, because it is not a separate subject, the environmental education appears in other subjects such as language, math, social studies, nature, and visual art (Palmer, 1998). As for Hong Kong, China, before the reunification, the concept of environmental education in Hong Kong has firstly appeared in Biology or Geography in the form of ecological research. Not until the 1980s has the environmental education been taken seriously. Later, in the white paper "pollution in Hong Kong: Action moment" which was an important milestone, Hong Kong's negligence of the environmental crisis was pointed out. After the 1990s, Hong Kong announced the "Guide of School Environmental Education" in 1992 and attentions of environmental issues began to be discussed in high schools (Palmer, 1998).

In 2004, China announced the "Environmental Education Implementation Guide (Pilot) in Primary and Secondary Schools". In this document, the environmental education was formally required to be implemented in the primary and secondary schools. However, the effectiveness of this policy is extremely limited. A report examining the problems of implementing school environmental education after the Guide was announced. The report surveyed 127 primary and secondary schools in 25 provinces in China, including 62 primary, 38 junior high, and 27 senior high schools. The report covered issues including environmental education curricula setting, funding, teaching materials, teachers and extra-curricular activities. The results of the survey showed that:

> Among 127 schools there were 111 that have set up (regardless of the form) environmental education programs, accounting for 87% of the total number of schools surveyed. Although the schools setting up the curriculum accounted for more than four fifths, almost none of these 111 schools set it as an official discipline nor did they include it into test subjects, so that it has no specific textbook nor sufficient funding. This made China's environmental education situated in a "thunder and no rain" distress condition (Jin, 2015).

Clearly, the "Ecological Civilization Construction" still insisted on the "reasonable consumer" argument, which, though being in line with China's general need of livelihood, is quite different from Chet Bowers' "pedagogy of ecological culture" that advocated minimizing the dependence of currency transactions. Since there is a tendency of "test leading teaching", China should make more fundamental ecological value reconstruction in environmental policy and the status of environmental education in schools, and take more positive actions and strategies to improve the ecological injustice situation more effectively.

4. Environmental Education Regulations and active eco-citizenship cultivation in Taiwan

Since the 1950s Taiwan began its industrialization; along with the economic growth, the industrial structure gradually transformed from agriculture into labor-intensive traditional industries. Industrialization on the one hand has attracted large numbers of youth from rural to urban areas who wished to earn a living or improve their lives. On the other hand, industrialization has brought about environmental pollution and destruction. During the 1970s, environmental problems have become really serious, such as groundwater pollution, heavy metal polluted farmland, large amounts of toxic gases being released by combusting metal waste, and untreated sewage being discharged by chemical plants. During the 1980s, civil societies began to organize anti-pollution and environmental groups, and environmental protest or movements like adequate housing, anti-nuclear, anti-reservoir, and tree conservation were quite active. Given the conflict between economic development and environmental protection, the "Environmental Protection Administration" of Taiwan was founded in 1987, in charge of the national air quality protection and noise control disputes, protection of water, waste management, environmental sanitation and poison management, environmental protection disputes, and environment monitoring. The local environmental protection bureaus of city governments in Taiwan were established progressively between 1988 and 1991, reinforcing the implementation of the local environmental protection work. As to formal regulations, based on the 1998 "National Environmental Protection Program", the "Environmental Basic Law" was published in 2002 to declare the aim of the policy of realizing ecological justice.

Article 1: This Act is formulated to raise the quality of the environment, advance the health and well-being of citizens, preserve environmental resources, and pursue sustainable development by promoting environmental protection. The regulations of other laws shall apply to those matters not regulated by this Act.

Article 2: ......"Sustainable development" means satisfying contemporary needs without sacrificing the ability of future generations to satisfy their needs.

Article 3: Economic, technological and social development shall equally emphasize environmental protection based on long-term national interests. However, in the event that economic, technological or social development has a seriously negative impact on the environment or endangers the environment, the protection of the environment shall prevail.

Article 5: Citizens shall uphold environmental protection concepts and lessen the environmental impact of daily life. In terms of consumer behavior, citizens shall, as a principle, practice green consumption. In terms of daily life, citizens shall carry out waste disposal reduction, separation and recycling. Citizens shall actively carry out environmental protection and be responsible for assisting the government in implementing measures related to environmental protection.

If compared with the "Ecological Civilization Construction " proposed by China, Taiwan emphasizes the ideal of sustainable development and intergenerational justice. It is not the "reasonable consumption" but the "green consumption" being underlined which reduce environmental problems. More importantly, when environmental protection and economic development are in conflict, "Environmental Basic Law" assures the priority of environmental protection and requires the public to fulfill their obligations to protect the environment.

As for the development of school environmental education, the Ministry of Education in Taiwan established the Environmental Protection Group in 1990 and by the next year, the "Environmental Education Committee". Yet the effectiveness of implementing environmental education in Taiwan is also limited. The key question could be too much emphasis on formal training in school teaching (particularly in primary and secondary schools). Within limited opportunities for education, teachers could only ask the students to follow the instructions, instead of guiding them to think about environmental awareness and actions to take. In the "Nine-year Curriculum Guideline" implemented in 2000, the environmental education is included as one of the important issues which schools are required to practice. The Curriculum Guideline listed five different dimensions of environment education, namely "environmental awareness and sensitivity to the environment", "conceptual knowledge of the environment", "environmental values and attitudes", "environmental action skills" and "environmental action experience". Although the people of Taiwan have been aware of the concept of environmental restoration, for instance, environmental preservation/reservation, conservation, protection and rehabilitation. Even a position of "eco-centrism" has been taken in the "Environmental Basic Law", however, the purpose of environmental conservation most people percept is inclined to "homo-centrism", which is still a long way from "eco-centrism".

Taiwan announced the "Environmental Education Act" in 2010 and became the third country in Asia that established statutory authority for environmental education act. The Act states the purpose of environment education as to enhance people's responsibility for protecting the environment and promoting social justice, in order to achieve sustainable development. To this end, the national legislation required all citizens, enterprises, government agencies and schools to attend at least four hours of environment education courses each year. In addition to school education within the system, in recent years, civil societies in Taiwan were encouraged to carry out experimental education. The educational content taught in this type of school is particularly inspired by movement of de-capitalist schooling. These schools are commonly against the form of schooling serving for capitalism and try to carry out the whole person education. Taking "Natural Way Experimental Educational Elementary School" (http://www.natural-way.com.tw/) for example, it does not aim to develop students' competencies and future employability, but to guide students learning traditional culture and slow culture through the meditation process of learning tea ceremony or calligraphy. In the process, students learn to dialogue to and gain knowledge of oneselves. In addition, the school curriculum also includes "Aboriginal Studies", in which students are to visit a mountain and a river as the season and climate change during the year. In the process, students are able to become humble when confronting the nature.

Perhaps the de-capitalist movement of the schooling reform in Taiwan may be demonstrated differently, but they still have values in common which, whether the abstract concept "harmony between heaven and human" or to guide students to learn the traditional wisdom, are closely related to the promotion of "ecological justice". Those schooling reformers may not know "eco-cultural education", but the teaching advocates have actually reverse the relationship between "high status knowledge" and "low status knowledge". In other words, valuable knowledge is not necessarily objective, rational or useful. It is necessary to be able to envisage the education of culture of local commonality.

> The relationships between the local cultural commons found in every community today and the industrial/consumer culture have not been mutually supportive. Indeed, the people who promote the expansion of the industrial/consumer dependent lifestyle, and thus the accumulation of capital, view the largely non-monetized cultural commons as potential markets to be exploited. (Bowers, 2008, p. 4)

According to Chet A. Bowers' argument, if environmental education curriculum is not based on cultural education and only consider the external forms of knowledge change, then the reform is only superficial. Neither the way of thinking of the students nor active ecological citizenship can be cultivated. Given the reason, the value highlighted by environmental education is possible to be transformed into practical actions by the students when we proceed from culture re-creation. "Ecological justice" can thus become a value that is truly supported. Considering this ideal, Taiwanese mainstream schools still need to put efforts on environment education.

5. Conclusion

As a Chinese ethic dominated society, China still insisted on one-party communist dictatorship politically after reform and opening up, while headed for capitalization in economy. As predicted by the world system theory, China's economic opening made it move closer to the world center. After the rise of the Chinese economy, however, some people became wealthy while a large portion of people and natural environment degraded and impaired. Confronting various ecological crisis, China put forward the policy of "ecological civilization construction" and in attempt to reinforce awareness of environmental protection and pollution control. Basically, "ecological civilization construction" is in fact a root metaphor of "progress", which deems the linear process of civilization as from an agricultural society to an industrial society and finally to ecologically civilized society. Therefore, changes embodying in the "progress" are linear, heading towards the opposite direction of the tradition (Bowers, 2005). China's current policy "ecological civilization" still demonstrates homo-centric theory, that is, the economy remains a priority followed by the position of environmental protection. In fact, this view is not only unable to solve problems derived from the basis of consumer relationship, but it will for certain become a kind of social Darwinism, which failed to see the great values of traditional ecological intelligence and local cultural common in solving ecological problems (Bowers, 2008).

Facing the conflict between economic development and ecological conservation, China proposed "ecological civilization construction" to develop its economy under the premise of reducing the environmental impair, and social justice is only demonstrated in compensation for the aggrieved. Compared to this, Taiwan after democratic reform for decades, human rights and ecological consciousness has become a common value. As for the ideals of the "Environmental Basic Law" and "Environmental Education Act" of Taiwan, proclaimed explicitly the purpose to reach ecological sustainability, that is, the priority of protecting ecology can not be overridden by the economic development. Once the line is crossed, they become unjust. Although the concept of ecology of Taiwan is aligned with the mainstream of international values, under the culture of emphasizing test scores and achievements, the environmental education in schools may not truly facilitate students' positive ecological citizenship. Fortunately, there have been a number of schools involved in the development of featured environmental curriculum, as well as a variety of environmental movement groups follow closely specific issues for the long term. There have even been many local experimental schools trying to place the local common culture in educational activities and to get rid of the clamp of capitalism. Such a de-capitalizing school education will certainly help to further realize the ideal of ecological justice.

❖ ❖ ❖

Chun Ping Wang is an associate professor of the Graduate School of Curriculum and Instructional Communications Technology at National Taipei University of Education (NTUE), where he is also dean of Center for Teacher Education & Careers Service. Dr. Wang served as the editor-in-chief of Journal of National Education Quarterly (2014.08-2015.12), serves as a member of Teacher Education Committee of Taiwan (2015-2017) and an executive editor of Journal of Comparative Education since 2015. He has studied extensively on philosophy of education, political philosophy, ecological ethics and teacher education. His current research focuses on Capabilities Approach theory and related educational issues.

References

Bowers, C. A. (2005). Educating for a sustainable future: Mediating between the commons and economic globalization. Retrieved from http://www.cabowers.net/pdf/Educatingforasustainablefuture.pdf

Bowers, C. A. (2008). Rethinking social justice issues within an eco-justice/ moral framework. Retrieved from http://www.cabowers.net/pdf/Social & Ecojustice.pdf

Bowers, C. A. (2011). *Perspectives on the ideas of Gregory Bateson, ecological intelligence, and educational reforms.* Eugene, OR: Eco-Justice Press.

Cai, Li-xia (2015). Curriculum Design and evaluation of environmental education based on the construction of ecological civilization. *China Population, Resources and Environment, 25*(11), 244-247. [蔡麗霞. (2015). 基於生態文明建設的環境教育校本課程設計與評價. 中國人口、資源和環境, *25*(11), 244-247.]

Chai, J. (2015). Chai Jing's review: Under the Dome – Investigating China's Smog. Retrieved from https://www.youtube.com/watch?v=T6X2uwlQGQM

Dong, Yan and Zhao, Ling (2015). On the Realization of Ecological Value of Justice. *Academic Exploration*, 1, 35-41. [董岩、趙玲. (2015). 論生態正義價值的實現. 學術探索, 1, 35-41.]

Hwang, Ruey-Chyi and Hwang, Chih-Tung (2007). The Three Problems of Environmental Justice Theory. *Taiwan Democracy Quarterly,* 4(2), 113-140. [黃瑞祺、黃之棟. (2007). 環境正義理論的問題點. 臺灣民主季刊, 4(2), 113-140.]

Jin, Xinrong (2015). Primary and secondary environmental education: Bigger thunder but little rain. *Environmental Education*, Z1, 5-9.[金鑫榮. (2015). 中小學環境教育:雷聲大、雨點小. 環境教育, Z1, 5-9.]

Liu, Ming Ming (2009). From "protection" to "reciprocation" : On the logic transmision of environmental legal obligation. *China Population, Resources and Environment, 19*(3), 46-49. [劉明明. (2009). 從「保護」到「回饋」－論環境法義務觀的邏輯嬗變. 中國人口、資源和環境, *19*(3), 46-49.]

Ministry of Education of People's Republic of China (2004). Guideline of environmental education in primary and secondary shcools. Retrieved from http://www.wwfchina.org/content/press/publication/eduFinal.pdf [中華人民共和國教育部. (2004). 中小學環境教育實施指南. Retrieved from http://www.wwfchina.org/content/press/publication/eduFinal.pdf]

Palmer, J. A. (1998). *Environmental education in the 21st century.* London and New York: Routledge.

Qian, Qiu-yue (2014) How ecological justice in contemporary China can be actualized: The theoretical Innovation since the 15th CPC Party Congress. *Journal of Nanjing CPC Institute*, 5, 78-82.[錢秋月. (2014). 生態正義在當代中國何以實現－兼論十五大以來黨的生態理論創新. 中共南京市委黨校學報, *5*, 78-82.]

Wallerstein, I. (2004). *World-systems Analysis: An introduction.* Durham and London: Duke University Press.

Wang, Chun-Ping (2008). Individualistic liberties, collective rights, global citizens: The problem of defining human rights and human rights education. Journal of Taiwan Normal University (Education), *52*(1), 25-44. [王俊斌. (2008). 個體自由、群體權利(與全球公民—論人權理念的發展與教育實踐. 師大學報(教育類), *52*(1), 25-44.)

Wang, Sheng-Hua (2013). Interpretation of ecological justice under the perspective of eco-Marxism. *Guizhou Normal University (Social Science Edition)*, 1, 38-41. [Wang. (2013). 生態馬克思主義視角下的生態正義詮釋. 貴州師范大學學報(社會科學版), *1*, 38-41.]

Wei, Sen-jie and Wei, Kwong-chi. (2008). International Perspective of ecological justice. *Journal of Chong Qing University of Science and Technology (Social Sciences)*, 1, 38-39. [魏森杰、魏廣志. (2008). 論國際視解下的生態正義. 重慶科技學院學報(社會科學報), *1*, 38-39.]

Wenz, P. S. (1988). *Environmental justice.* Albany, NY: State University of New York.

Wikipedia. (2016). List of countries by GDP (PPP). Retrieved from https://en.wikipedia.org/wiki/List_of_countries_by_GDP_(PPP)

Zhang, Binquan (2009). Ecological civilization and ecological justice. *Journal of Central South University of Forestry and Technology (Social Sciences)*, *3*(1), 10-12. [張兵權. (2009). 生態文明與生態正義. 中國林業科技大學學報(社會科學報), *3*(1), 10-12.]

How the Technology of Print Promotes Abstract Thinking

by Chet Bowers

Ironically, the two most basic differences in how we experience the openness of the spoken word and the initial certainties that many readers derive from the printed word, are not given serious attention at any level of formal education. The printed word, from the earliest years, is represented as a more advanced and thus as more progressive and enlightened than reliance on the spoken word. This status difference is captured in the metaphors of literacy and illiteracy, as though the latter is a badge of dishonor and backwardness.

As print technology was first introduced centuries ago to record the shipment of goods, it has become a powerful, indeed indispensible, influence in advancing knowledge and providing ways for people to share their ideas beyond the reach of the traditionalist and reactionary forces that stand guard in too many communities. The uses of print have often opened doors of imagination and deep reflection often missing in oral communities where conformist/authoritarian thinking has taken over. While I am inclined to view authoritarianism as more likely to occur in print-based cultures, which I will explain more fully, there are deep personality issues that will never be fully understood by the cognitive scientists with their brain-centric reliance upon what MRI brain scans reveal. The continuing narrowing of life forces to what is occurring in brain, which excludes the ecology of communication occurring in all natural and cultural systems, continues to be shaped by the legacy of the 16th and 17th century.

Nietzsche, more than anyone else, understood the connections between the deep largely unconscious psychology of authoritarianism, which he understood as the more destructive expression of the will to power, and living in the world of becoming where the will to power is expressed in creativity and openness. Before examining more closely how print misrepresents the world in which we live, several of Nietzsche's insights need to be noted as they will help avoid adopting simple causal relationships that exclude the inner forces behind people's behavior. These include the following: "Knowledge works as a tool of power (1968, 266); "It is our needs that interpret the world; our drives and their For and Against. Every drive is a kind of lust to rule; each one has its perspective that it would like to compel all other drives to accept as a norm." (267); "Everything simple is merely imaginary, is not 'True." But whatever is real, whatever is true, is neither one nor even reducible to one" (291); First proposition. The easier mode of thought conquers the harder mode; as dogma…. to suppose that clarity prove anything about truth is perfect childishness. Second proposition: The doctrine of being, of things, of all sorts of fixed unities is a hundred times easier than the doctrine of becoming, of development—. " (291). And finally, "…every elevation of man brings with it the overcoming of narrower interpretations; that every strengthening and increase of power opens up new perspectives and means believing in new horizons…" (330). Nietzsche ends this passage by claiming that there is "no truth". As I will argue, there is no objective knowledge and data. So what is there in the taken for granted linguistic constructions of what is taken to be reality that leads to these fateful misconceptions?

Before turning to consider the characteristic of print that have been largely ignored because of the history of considering print as the technology essential to human advancement it is important to make several other observations that will help avoid placing print and data in the dichotomous world of good and evil. Taking account of different contexts (such as considering the impact of print on tradition-oriented cultures) and relationships (such as how it affects the exercise of ecological intelligence) needs to be taken into account in assessing when its use becomes an ecologically destructive force. With the life threatening changes in the world's ecological systems, as well as the massive threats to personal security and privacy—including the global threats to the nation's infrastructure and institutions that the computer culture continues to represent as the latest expression of progress, it is essential that relational thinking is part of the following discussion.

A superficial knowledge of other cultures, as well as one's own—including the thinking that led to important advances in knowledge are also part of why print has been so valued over the centuries. Gifted writers have been able to use the printed word as a mirror that enables us to see ourselves, and our conceptual and moral double binds, more clearly. And without print we would be limited to a barter economy, and to the uses of technologies that are learned through face to face mentoring relationships. These advantages are largely taken for granted by everyone who has graduated from higher education. But it is this one-sided view of print that has led classroom teachers and university professors to ignore engaging students in a serious and in-depth discussion of how print, for all its many uses, undermines the exercise of ecological intelligence that comes so naturally in many oral communities—yet remains a challenge even in our face to face interactions that too often reproduce the abstract thinking acquired from an over reliance on print-based cultural storage and communication.

Deconstructing Print-Based Cultural Storage and Communication:

One of the paradoxes we now face, given the rapidly deepening ecological crisis that is leading to the endgame of widespread social chaos as billions of people struggle to survive the shortage of water, protein, and the complete breakdown of moral norms of which hackers are the now the leading edge, is the growing influence on globalizing the mode of communication that marginalizes awareness of local contexts. Like all technologies that amplify and reduce cultural patterns that, in turn impact human experience, print and data have the following inherent limitations.

- **Print can only provide a surface knowledge of ideas, events, and processes.** Print is unable to represent of the multiple influences and relationships within the cultural and natural ecologies that constitute local contexts. The writer too often lacks a knowledge of the semiotic processes that make the emergent and relational nature of the information rich natural and cultural ecologies. Even if the writer were aware of the tacit and taken for granted cultural patterns she/he would not be able to fully translate them into print. Thus, what print provides is only a surface knowledge, which may be useful in certain situations. The surface knowledge, in turn, provides only an abstract understanding that too often becomes the basis of what is communicated to others. Twitter, texting, e-mails, and cell phone communications further reinforce the tradition rooted in centuries of print-based cultural storage and communication that held that abstract knowledge has higher status and is more efficient in guiding behavior and social policies than knowledge informed by what is being semiotically communicated through the emergent, relational, and co-dependent natural and cultural ecological systems—which is the world we interact with on the streets and other public spaces.

- **What is committed to print immediately becomes outdated in the ecological world of constantly changing relationships and multiple levels of message exchanges:** What is committed to print can be constantly updated, but it can never reproduce the dynamic nature of ecologies which carry forward influences from the past and are constantly changing. To test out this generalization, give a printed account of an ongoing conversation or the action of ocean waves, and then assess what was omitted in a printed account. Both are examples of the changes in patterns, the multiple influences on these patterns, and the inability to represent them fully in print except in a highly abstracted and static manner. The problems connected with print-based cultural storage and communication are further compounded when readers impose on what is written their culturally influenced interpretative frameworks, which often lead to another level of abstract understanding or becomes entirely distorted by the reader's taken for granted interpretive framework. As the level of abstraction becomes increasingly removed from the patterns and messages of living ecological systems, there is an increasing loss of accountability for what the printed word represents.

- **The abstract thinking fostered by print too often becomes interpreted as having a universal meaning.** In being able to represent only a surface knowledge of events, ideas, and so forth, as well as its dated representation of what ecological systems have already moved beyond, the printed word marginalizes awareness of the hidden cultural influences on what the writer thinks is important. This process has become especially acute in the West where print has become the basis of high-status knowledge, and will become even more so as people increasingly find it more convenient to engage in digitally mediated communication. As we are witnessing in how digital communication is changing patterns of thinking, including attention spans, the historical influences on what is becoming only surface knowledge is increasingly slipping into the realm of silence. Without this historical knowledge of the forces shaping the present, turning abstractions into universal truths become easier for both the writer and the reader.

- **Print reinforces the conduit view of language that undermines awareness that words are metaphors that carry forward earlier cultural assumptions.** Print simply serves to hide the misconceptions and hubris of the writer by creating the illusion of objectivity and factualness. The conduit view of language as a sender/receiver process of communicating objective data and knowledge further undermines the reader's awareness that words have a history, that most are metaphors whose meanings were framed in earlier eras when there was no awareness of the limitations of ethnocentric thinking and that there are ecological limits to the western approach to progress. These understandings should be part of the formal education of everyone, including environmental scientists, computer engineers, and the people who write the software and create the video games. The conduit view of language can be addressed by introducing phrases and information that foster an awareness that the meaning of words has a history, and that their meanings can be reframed by adopting ecologically and culturally informed analogies. (Bowers, 2012, 107-166)

- **Print reinforces the western myth of the autonomous individual who relies primarily upon a visual relationship with the external world.** As most patterns of communication in the West do not make explicit that words are metaphors whose meanings were mostly framed by analogs chosen in the past, most individuals assume they are making independent judgments about the events, ideas, and so forth represented in print. The static yet constantly updated world of print is profoundly different from that of oral cultures where participation leaves less room for standing back as a supposed external observer who makes objective and critical judgments. The printed text also introduces an asymmetrical power relationship between the writer and the reader, with the reader (or teacher and professor) reproducing this same asymmetrical power relationship when conveying to students what was in the printed text. This power relationship also reinforces the illusion of being a rational and self-directing individual. Again, the failure to introduce students to how being socialized within culture's largely taken for granted languaging processes, which computer-mediated learning is unlikely to address, perpetuates the misconception that rational thought, critical inquiry, and thus individual judgment are free of cultural influences.

- **Print is inherently ethnocentric.** Orality involves responding to multiple relationships, influences past and current, messages both tacit and explicit, and what constitutes the appropriate moral behavior in a world of changing relationships. Print, which has been the dominant technology in colonizing other cultures through the creation of maps, written treaties and contracts, as well as linguistic/ideological impositions, has led to treating oral cultures as backward, illiterate, and thus easily exploitable. The current promotion of digital technologies in predominately oral cultures is a continuation of the colonizing agenda of bringing these cultures into the realm of literacy and a market economy that relies upon individualism and the destruction of the local cultural commons. At the most basic level, print cannot encode the multiple messages communicated in the living moment, and within and between the cultural and natural ecologies. The cultural practice of relying upon all of the senses, memory, and the openness to negotiating a change in meanings as speakers speak, listen, and read the body language of each other is more often found in oral cultures. Tacit knowledge also plays an important role in oral cultures as well as a metaphorical language that is influenced by changes occurring within the interacting ecologies.

- **Most writers and readers are unaware of the taken for granted cultural assumptions that influence their interpretations of the world that take on the appearance of objectivity when encoded in the printed word.**

So how did the West take the turn way from the face to face and co-dependent mentoring lifestyles that kept starvation at bay by giving close attention to the cycles of emergent and relational changes occurring within their bioregion? There are many explanations, with most reflecting the modern mindset of identifying the adoption of new technologies such as the printing press, the scientific method, the agricultural revolution that replaced the scratch plow with the moldboard plow that had a metal cutting edge that turned the soil over—thus increasing crop yields. The increase in population centers lead to the need to rely upon the printed word for understanding ongoing events and for reading about one's place in the universe. It is important to note that the early reliance upon the printed word as did not take account either of the emergent, relational, and co-dependent nature of the life supporting natural and cultural ecologies, or that the human relationship in these ecologies involved histori-cally layered patterns of interpretation.

These silences were reinforced by the elevation of abstract thinkers to elite status. Those who mentored others in how to grow and prepare food, in the crafts and uses of community-scaled technologies, and in the patterns of mutual support within com-munities, relied upon face to face communication— and not on the printed word. Generally overlooked today, is the role western philosophers and social theorists played in establishing not only what constituted high status knowledge, but also the abstract ideas that continue to underlie the misconceptions that are now taken for granted by the current elites whose agenda is to digitize and thus to further expand the consumer-oriented culture now being globalized as the model of human prog-ress. If one can avoid becoming caught up in the tribal debates between the western philosophers and the efforts of later generations of philosophy professors to promote this earlier tribal model of rationality among their students, it becomes possible to recognize other aspects of their legacy that now limit our ability to alter the ecologi-cally destructive pathway we are now on.

Western philosophers were, with only a few exceptions, abstract thinkers who relied upon print for communicating with other abstract and ethnocentric thinkers about adopting their culturally uniformed agendas for guiding future social develop-ment. Plato's arguments against oral narratives, John Locke's justification for deter-mining the ownership of private property, Adam Smith's discovery that there is such a thing as free markets and an invisible hand that ensures the survival of the most competitive in trucking, bartering and trading, René Descartes's insight that we can live a more rational existence if we ignore traditional sources of knowledge and view ourselves as separate from the world we think about and act upon, Roger Bacon's advocacy of using science to control nature for human benefit, John Stewart Mill's augment that everything should be questioned, and Herbert Spencer's discovery that all life is subject to Nature's law that ensures the survival of the fittest, John Dewey's pronouncements on the superiority of scientific/experimental inquiry over cultures dominated by "savage" (his term) and spectator approaches to knowledge. Not to be outdone in the realm of abstract and culturally uniformed thinking are Ayn Rand, Milton Freidman, and E.O Wilson. The latter argued that the brain is a machine---a problem in engineering. He also claimed that scientists should pass judgment on which moral and religious sentiments people should live by.

As we enter the world being transformed by the digital revolution we find com-puter scientists such as Ray Kurzweil, Hans Moravec, and Gregory Stock carrying forward the Titanic mindset of these earlier elites who not only ignored the absolute dependence of humans on the viability of natural systems, other cultural systems of knowledge, and an understanding of the cultural traditions that should be intergen-erationally renewed.

Today' current Orwellian use of our political metaphors can be traced to the ideas of this long-history of abstract theory and thinking. That is, abstract thinking is free of being held accountable for the variations, complexities, histories, and moral norms that exist in local face to face cultural contexts—which should not be romanticized. Many of the narratives that named the West as an advanced civilization also promoted gender, racial, and other prejudices, especially toward oral and tradition-oriented cultures.

Especially problematic is that many of today's leading politicians, economists, and now nihilistic populists activists have embraced the ideas of Ayn Rand who argued in The Virtue of Selfishness (1964) that individuals should exercise their rationality in order to achieve the most personal happiness and self interest. Her own abstract approach to rationality led her to claim that governments should not provide safety nets for those in poverty, that altruism and empathy toward others were values promoted by the weak, and that governments have no right to tax what others had earned, carries forward the abstract thinking shared by today's capitalists who ignore that we live in a relational and co-dependent world that still includes the natural ecologies that are fast disappearing.

The irony is that these core libertarian ideas are widely referred to as conservative, while people concerned with conserving habitats, species, social justice achievements, linguistic and cultural ways of knowing, and the diversity of the world' cultural commons are identified as liberals. How this Orwellian misuse of our political language relates to the unconscious influence that print has had on the lack of accountability for how our political vocabulary is being used can be seen in how the abstract words such as conservative, liberal, tradition, emancipation, progress, indigenous, and so forth are being used as free floating labels that have no relationship to the cultural patterns of everyday life whose complexities cannot be adequately represented by these labels. If we were to do an ethnography—that is, a careful description of the lived cultural patterns that fit the core beliefs and assumptions historically associated with these different political metaphors— we would find that these abstract words fail to account for the range of traditions that people take for granted, the ways they struggle to reconcile cultural norms others take for granted with their need to challenge how their own lives are being constrained, and the complexity of their relationship with the natural environment.

Each individual's life is also characterized as emergent, relational, and co-dependent upon the information (semiotic) networks and support/constraints systems of the natural and cultural ecologies within which they live. The abstract use of political labels, like the abstract nature of the printed word, ignores the complexities of the ecological connections with the past, with others, and with the possibility of ecological collapse as 9 billion people increase the demands on a environment whose capacity of self renewal is in rapid decline.

Summary of what is missing in how print is understood:

The following will help minimize the "immaculate conception"
way of thinking that print provides objective facts and information about a world
that supposedly is free of human interpretation.

- To reiterate: what appears in print always has a human authorship, which means that it represents a culturally influenced interpretation.

- Writers and readers, even the most gifted and thoughtful, cannot be fully be aware of the earlier patterns of thinking that are encoded in the metaphorical language they largely take for granted.

- Readers, who bring their own unresolved (perhaps not even recognized) psychological issues and taken for granted conceptual world to what they read, are part of yet another layer interpretation—and misinterpretation that may further imperil our future. Those in denial about the ecological crisis are examples of the latter.

- Writers are generally unaware of how print, whether in books or on the computer screen, makes it difficult to recognize that words have a history and thus reproduce earlier expressions of intelligence that encode the taken for granted assumptions of earlier eras.

- Most writers and readers take for granted that they are autonomous thinkers and thus are giving an objective account of that part of the world they write and read about.

- The inner psychological forces Nietzsche identified with the will to power that can also be referred to as hubris, which also includes the power play he referred to as ressentiment, are also aspects of the ecology of writing and reading.

- As the cultural, linguistic, and psychological influences on what is written are often hidden from readers, the printed word may be interpreted as stating an objective and universal truth. That print can never fully reproduce the multi-layered emergent and relational nature of the ecologies we misrepresent by referring to "contexts", which is another well intended abstractions, contributes to the sense of certainty valued by most writers and readers.

The main difference between print and data:

Because data is a construction of the scientific mindset that claims the mythical powers of obtaining objective knowledge free of all cultural/linguistic influences, data now serves a number of important functions in undermining cultural traditions that have not been entirely colonized by the radically reduced conceptual world of the computer scientists, cognitive scientists, and libertarian/market liberals. The influences that operate in the ecology of print also operate in the world of data, and if understood would challenge the high status now accorded to data.

The determination of what to collect as data is always based on someone's interpretation— which usually means someone's will to power that reflects being socialized to take for granted the explanatory power of the root metaphors of mechanism, and its supporting root metaphors such as anthropocentrism, individualism, economism, and progress. The supporting vocabularies of these root metaphors (interpretive frameworks) exclude the vocabularies essential to understanding the diversity of cultural histories, their different approaches to ecological knowledge and daily practices, and their moral ecologies and wisdom traditions. The metaphor of "objective" also hides how the collection and use of data undermines awareness of the powerful role it plays in achieving different ideological ends. In the hands of environmental scientists, it serves to justify practices that conserve species and habitats, and to challenge the environmentally destructive practices of market forces and government policies that reflect the interests of powerful elites.

When used by libertarian and market liberal groups, which includes the rapidly expanding digital culture, the idea of objective data supports the current ideology that seeks to replace humans with algorithms and robots—and to bring all aspects of organic life, including humans, under the control of massively connected computer systems now being justified on the grounds that this form of progress needs to be understood as how the process of evolution is on the cusp of replacing organic life with super intelligent computers.

Whether used by environmental scientists and groups working to achieve a more socially just and ecologically sustainable cultures or by those still under the control of the cultural myths that continue as the main legacy of the abstract thinking philosophers and social theorists who are now leading us down the path to a techno-fascist future, data is always interpreted. What is particularly difficult to grasp is why so many people now understand that nearly every aspect of their lives is being is electronically monitored and stored as data, which then is use by strangers whose values and political agendas are unknown except for how they are making the lives of more people economically insecure and subject to constant harassment by strangers promoting scams and the latest consumer opportunity. Is this because they have been conditioned by the legacy of the abstract thinking philosophers, social theorists and religious leaders to ignore their traditions of more face to face community-centered and thus less consumer driven lives?

References

Bowers, C. 2012. *The Way Forward: Educational Reforms that Focus on the Cultural Commons and the Linguistic Roots of the Ecological /Cultural Crises*. Eugene, OR." Eco-Justice Press.

Nietzsche, F. 1968. *The Will to Power*. Edited by Walter Kaufmann. New York: Vintage Press.

Rand, A. 1961. *The Virtue of Selfishness*. New York: Signet.

After-word

COMMENTS ON THE CURRENT STATE OF ECO-JUSTICE THINKING

by Chet Bowers

The papers in this volume provide a mixed account of the relevance of eco-justice thinking as an alternative to the paradigm that climate change has now brought into question. Unfortunately, none of the papers provide a comprehensive overview of the issues that should be the focus for understanding that we are not autonomous individuals but live in natural and cultural ecologies—and how this understanding can lead to more socially just community-centered lifestyles. Before discussing what is missing in all of the papers, it is necessary to identify the insights in several of the papers, as well as lines of thinking that have the potential of making a genuine contribution that future generations will need to develop further. This is, after all, the real test of whether the current state of thinking will be recognized as avoiding the misconceptions rooted in the distant past.

The paper by Thomas Nelson and John Cassell presents the evidence that should bring a sense of urgency, a wake-up call if you will, to identifying the sustainability insights that later generations can build upon—that is, if the growing awareness of the changes resulting from the degradation of natural systems (ecologies) does not lead to the chaos that can be exploited by right-wing politicians who set out to perpetuate the old neo-liberal consumer-dependent industrial system. The advantage of such a political take-over is that most Americans already take for granted the deep cultural assumptions the underlie the episteme of classical and now neo-liberalism.

Nelson and Cassell make the point that what is needed is a different paradigm or interpretative framework not based on the current misconceptions about individualism, a human-centered world, and the false ontology that promises infinite progress. The paradigm shift is to be based on systems thinking, which might be better understood as a world characterized as emergent, relational, and co-dependent.

Nelson and Cassell take their readers to the edge of an important break through, which is that in this relational world described by Gregory Bateson where there are no isolated, autonomous and fixed entities, but only differences which make a difference. (Bowers, 2011) These differences which make a difference are the basis of the information exchanges in this emergent, relational, and co-dependent world of natural and cultural ecologies. Unfortunately, they do not explore the importance of Bateson's ideas to making this paradigm shift. If they had explored Bateson's ideas they would have found that the many ecological relationships are actually the pathways of communication we are now beginning to understand as semiotic exchanges that reflect the patterns of communication characteristics of different forms of life—from the most simple organisms to the more complex.

The focus on the patterns within natural and cultural ecologies that connect and transform each other leads to recognizing another point that Nelson and Cassell make: that is, linking learning to understanding the everyday cultural patterns or communication networks of everyday existence (what is communicated in the emergent relationships) that we have more traditionally referred to as cultures. This is where Nelson and Cassell needed to expand on just how different this is from learning from the printed word—which is, as I will explain in the last essay, the basis of the abstractions that encode many of the misconceptions handed down from the past. The pervasiveness of abstract thinking can now be seen in the shift away from face to face, context specific learning which is more likely to occur outside of the classroom, and toward more computer mediated, print-base learning that reinforces abstract thinking that marginalizes awareness of the information exchanges occurring in local contexts. Students need to recognize that these information exchanges vary according to differences in the cultural patterns that connect.

There are many shared silences in these papers, and one of the most surprising is that none provide a discussion, or even a mention, of the digital revolution that is changing the basis of memory itself, further marginalizing the possibility of democratic decision making, expanding the ability of corporations to reach deep into human consciousness in ways that exploits the culturally conditioned proclivity to be a consumer, and represents the marriage of scientism, technology, and corporate values as the latest expression of the evolutionary process itself. (Bowers, 2000; 2014; 2016) This oversight is critical to all of the papers, but especially the papers by Joseph Progler and Rolf Jucker—but for different reasons.

Progler's focus is on the relational nature of existence, which he interprets through the lens of how the various academic disciplines are based on understanding the relationships within and between the different facets of culture that represent the discipline's conceptual territory. Each discipline—history, sociology, anthropology, religion, and so forth, provide examples of relational thinking. But Progler never moves the discussion to where the ontological issues are confronted as to whether this is really a world of autonomous, discrete entities that can be represented as objective facts and data—or the one constructed by earlier social theorists who relied on the authority of print while ignoring the emergent, relational, and co-dependent world of everyday experience. Depending upon the organizing narratives, the world of things and event can be studied in terms of their relationships with other autonomous and discrete events and entities, but this is profoundly different from the ontological understanding that there is no permanence in the world—but only in the abstractions that most Westerners assume represents reality such as an economic system, the autonomous individual, freedom, and so forth. Again, what is missing in Progler's focus on the relational thinking promoted in the disciplines is a discussion of what careful attention to daily experience reveals: namely, that impermanence is the dominant reality in both the natural and cultural ecologies. While the long tradition of print based representations of reality promote abstract thinking, where the abstractions are taken as real and thus seldom examined in terms of local face to face and semiotically rich contexts, the ontological reality is that we live in an emergent, relational, and co-dependent world where relationships are the communication pathways where, to quote Bateson again, "a 'bit of information is definable as a difference which makes a difference" (Bateson, 1972, p. 315) The behavior of the individual, a plant that is being pollinated, changes in the chemistry of the ocean, all involve introducing differences that are the sources of information for others that share the same semiotic patterns of communication, which may be chemical, differences in temperature, body language, the spoken word that is not understood by the Other.

The emergent, relational, and co-dependent nature of the living natural and cultural ecologies can be seen the patterns, of everyday conversations, in the behavior of animals and even in plants. If Progler had been less focused on explaining the many ways that the academic disciplines and religious lifestyles promote relational thinking, his essay could have easily led to deepening our understanding of ecologies. That the potential explanatory power of eco-justice only makes sense within a world understood as natural and cultural ecologies that are now signaling that we are entering a period of extreme crisis never before experienced by humans, we need to ask another question about whether the relational thinking promoted in academic disciplines enabled the members of these disciplines to promote awareness of climate change and to address all the other evidence of environmental degradation.

It was initially a struggle for faculty in the various disciplines to have their research on environmental issues taken seriously by colleagues trapped in the paradigm that had shaped the discipline. This has now changed, but we still find that the awareness of climate change has not led to radical changes in the paradigms that still dominate the various academic disciplines. Relational thinking is important, perhaps inevitable given the emergent and co-dependent nature of the world, but its potential to awaken people to what some scientists are now seeing as evidence of the sixth extinction of life on this earth is undermined when the academic disciplines carry forward the silences and misconceptions left us by western philosophers and social theorists. (Kolbert, 2014). What will the next generation, the one now being indoctrinated by the digital revolution to accept the authority of data and the idea of progress that promises to replace humans with super-intelligent machines, take away from Rolf Jucker's five point summary of what is needed if addressing the ecological crisis is to be in the hands of scientists? Unlike his earlier writings, he now seems focused on rescuing the environmental movement from the mindlessness of what he terms of "medieval mindsets" that are expressed in the idea of the cultural commons, and a concern about traditions passed forward through face to face intergenerational communication and mentoring in how to live less consumer dependent lives. As both Wendell Berry and I point out, scientists have made important contributions in a wide variety of fields, but they have a proclivity for not recognizing the limitations of the scientific method, which is often expressed in their efforts to overturn traditions by introducing what they regard as the new technologies that will give them more control of the environment—including people. (Berry, 2000; Bowers, 2016)

The science that Jucker is promoting as the one true way of eliminating beliefs from the past, the influence of religions (how this can be done where the religion and cultural beliefs such as in Eastern cultures are inseparable is not explained), as well all external influences on an individual's rational capacity, is the science that has introduced toxic chemicals into every aspect of our bodies and environments, engaged in doing lobotomies on helpless individuals as well as promoting the eugenics movement, promoted an ethnocentric view of intelligence, created more lethal weapons systems, and now is promoting the development of digital technologies that is undermining democratic decision making and such traditions as personal privacy and security from nihilists who hack and engage in cyber attacks.

I agree with Berry that the destructive nature of western science probably balances with all the genuine achievements of modern science. But this is not the view of science that is in Jucker's five point agenda. There is no sense of awareness that scientists often cross beyond what their epistemology can account for, and engage in scientism where their actions impact aspects of culture they do not understand. This is clearly evident in what the science-based digital revolution is now overturning. Where are the voices of scientists protesting how leading computer scientists now claim we are entering the era of a post-biological future where computers will increasingly replace humans in the work place, in decision making in fields ranging from simple work procedures to making decisions affecting larger ecological and cultural systems. Is the digital revolution driven by an ideology that scientists take for granted or is it under the control of the natural forces that Darwin described? Given that the every aspect of the industrial revolution depended upon scientific advances, including the sciences that explained how to shape and control people's propensity to consume, we should recognize that scientists are not uninfluenced by the misconceptions handed down from the past that are encoded in the vocabularies that are inherited—and not created by autonomous rational individuals. Scientist have also been socialized to think in this inherited vocabulary, and the word that they still have not understood as based on the misconception of the Enlightenment thinkers is "progress" which led to an oversimplified understanding of traditions—which Jucker repeats.

Jucker's argument that we need to rely upon our own mind "without guidance from others" is especially problematic in a paper that purports to eliminate the romanticism from thinking about lifestyles that contribute to an ecologically sustainable future. I'm aware that the books I published with the Eco-Justice Press have been largely ignored, but I thought that the many chapters explaining how the metaphorical nature of most of our vocabulary carries forward the silences and misconceptions of earlier eras, which means that most of our ideas reproduce earlier misconceptions which we mistakenly assume to be our original ideas, might have led to rethinking the myths surrounding the ideas of rational thought and objective knowledge. The ideas were boiled down to their most basic formulation, such as recognizing the words have a history, and that their meanings change as new analogs are introduced that reframe the meaning of word. And that a person's ideas are, in part dependent upon the taken for granted use of an inherited vocabulary—which I have described as part of the process of the linguistic colonization of the present by the past. (Bowers, 2011, 69-92; 2012, 15-26, 107-166; 2013, 21-40)

Recognizing the we live in an interpreted world has huge implications for how we educate the generation that will face the chaos that will accompany the shortage of water as glaciers disappear, and as sources of protein are affected by the acidification and overfishing of the oceans, and species and habitats disappear under the pressure of a world population that is predicated to exceed 10 billion by the end of the century. There are other questions to be raised by his recommendation that everyone should use their own minds rather than be influenced by others. One that stands out in terms of the support now given to Donald Trump is whether Trump's hard core supporters are examples of individuals who mistakenly assume they rely upon their own minds.

Jucker's five point agenda for turning over the rational process to scientists is especially problematic when we consider the history of scientists such as in Nazi Germany where social Darwinism guided their research. Scientists in this country have also strayed into the Alice and Wonderland of scientism, such as when E. O. Wilson refers to the brain as a machine and a problem in engineering and Lee Silver suggests in *Remaking Eden* (1997) that scientists should engineer a new species of gene-rich individuals who will take control of the cognitive functions in society, and with the "naturals" taking on the physical tasks. Given Jucker's proclivity to see only the positive contributions of scientists, and there are many, he brings this same oversimplifying mindset to seeing traditions as sources of backwardness, and thus overlooks that the complexity of traditions re-enacted and modified in everything we do, such as in the traditions that separate writing from speaking. The traditions carried forward by the digital revolution, with the emphasis on the printed word and the myth of objective knowledge that is encoded in the idea of data, are undermining the traditions that carry forward many of the social justice achievements of the past. What few past and current Enlightenment thinkers recognize is that all social justice gains, in being carried forward over four generations, become traditions—such as not supporting slavery, not exploiting child labor, and so forth.

In spite of what is problematic about Jucker's paper, he does make several important points about not embracing the romantic thinking that too easily accompanies ideas that have merit, such as the idea of the cultural commons—which are so complex and may encompass social practices that are racists and exploitive in others ways. On this point he needs to be taken seriously, but not to the extent that what is especially important about the diversity of the world's cultural commons is that carry forward knowledge of how to live less consumer dependent lives gets lost sight of. This is the feature shared by most cultural commons, as well as provide for the sharing and development of community-centered skills that will become even more important as the digital revolution leads to higher levels of unemployment. Jucker's warning can be taken seriously by adopting an ethnographic approach to identifying the social justice and injustice characteristics of different cultural commons.

The paper by Ethan Lowenstein and Nigora Erkaeva makes an important contribution to eco-justice thinking by focusing on the community building characteristics of the Southeast Michigan Stewardship Coalition. It provides an example of what collaboration among a wide variety of community groups, including teachers, can achieve when the conceptual underpinning are clearly understood. Lowenstein and Erkaeva also explain how theory, such as the ideas about the merits of place based education, can be used as the basis for community renewal. The discussion of how the coalition reform effort differed from the dominant model they refer to as the "unilateral partnership" where individual organizations operate according to the old criteria of profits and losses, and meeting predefined measurable goals, is especially useful as they explain how language can dictate both the process and outcome. They point out that the emphasis of the Stewardship Coalition was to promote transformational partnerships, which represented a clear break from the capitalist mindset.

Students reading their paper could use it as a model for understanding the role that language plays in the process of social transformation. That is, the differences between how the language that supports a conceptual framework, such as a world that is emergent, relational, and co-dependent, or a world dependent upon market forces and status differences, takes account of cultural impacts on the viability of nature systems. What their paper brings out is the primacy of the language processes that are too often overlooked due to how the reality shaping role of language is too often taken for granted without anyone asking whose history is being re-enacted here. Not only are the earlier misconceptions and silences encoded in today's meaning of words, which together serve as powerful interpretive frameworks we now understand as ideologies, but they also influence whether thinking is dominated by abstractions or lead to giving careful attention to what is being communicated through the face to face relationships—which leads to different levels of interpersonal accountability which is too often missing in the abstract print dominated world. These are key issues that need to be addressed in moving eco-justice thinking forward.

Chung-Ping Wang's essay on The Two Faces of Eco-Justice in Chinese Society, is from a North American perspective ground breaking in that it raises questions that are not currently being considered by educational reformers concerned with eco-justice reforms. Given the differences in cultural traditions, where in North America there is the tradition of the autonomous individual and resistance to policies announced by a central governmental authority, particularly useful is Wang's description of how both China and Taiwan have adopted laws that emphasize the need for curriculum reforms that promote living in harmony with nature. The environmental laws passed during the Nixon administration, while limiting the degradation of natural systems in the United States, stopped short of promoting educational reforms that required students to practice "green consumption" or to consider what living in harmony with nature would entail. In the case of Taiwan, the Environmental Education Act of 2010, according to Wang, involved a shift away from promoting the values and skills that further a capitalist economic system, but instead promoted learning the intergenerational traditions that had not been monetized, and that have a smaller impact on natural systems—what some in North American have been referring to as the cultural commons. What Wang's essay raises is the question of whether it is possible, given how our political system is controlled by corporations and politicians who possess the libertarian/liberal mindset, for a national reform effort to appear that promotes educational reforms that stress non-consumer values and ways of thinking. The Taiwan reform of 2010 appears as the exact opposite of the Common Core Curriculum reforms promoted by American corporations that want to ensure a steady supply of school graduates to fill the need for workers in the industrial system that is being transformed by the digital revolution.

Wang's analysis of the differences between the Chinese and Taiwan approaches to promoting educational reforms highlights the tensions and contradictions in both systems. One of the problems that Wang mentions is the differences between the intent of Taiwan's Environmental Education Act and its actual impact on what is being learned in the classroom. Requiring that all students spend a certain number of hours addressing environmental issues is a first step, but like so many reforms mandated by governments, it falls short of taking account of how the values of the dominant consumer oriented culture continues to influence consciousness. What is missing is an actual curriculum that encourages students to give close attention to the many ways that consumerism and the western values of individualism and technology are encoded in the daily cultural patterns that students otherwise take for granted. This would include a deeper understanding of how consumerism, and the advertising and digital systems that promote it, represent an alternative vocabulary to that of the low status forms of traditional knowledge and skills passed forward through face to face relationships within the community. That Taiwan educational reformers are recognizing that challenging the high status knowledge of the capitalist culture is essential to eco-justice educational reforms is a very positive development—one that western educators can learn from.

As Wang points out, the Chinese emphasis on reforms that foster an "ecological civilization" are based on "homo-centric" assumptions, and that progress measured in making the transition from an agricultural to an industrial society, involves a basic contradiction that it still leads to overshooting the ability to live within environmental limits. Wang's discussion of how traditional tensions between rural and urban, which also exist in North America, affect national efforts to promote environmental education suggests what has been missing in current eco-justice thinking in North America. But there is another difference that separates the main foci of the Chinese and Taiwanese approaches to promoting environmental educational reforms, which is that the indigenous cultures across North America have a long history of learning from the environment in ways that led to the development of ecological intelligence that understands the world as emergent, relational, and co-dependent—which requires giving attention to the semiotic systems of communication within the natural and cultural ecologies. Learning from giving close attention to the communication occurring in the natural and cultural ecologies requires yet another reversal between high-status knowledge (which is largely dependent upon the technologies of print and data) and the low-status knowledge acquired from the diverse semiotic systems operating in both the natural and cultural ecologies.

Overall, Wang's essay moves the discussion of eco-justice issues into new territories marked by deep cultural and ideological differences, while at the same time highlighting areas of common concern that deserve further elaboration—such as how reforms in the classroom can lead to a deeper understanding of the complexity of lived cultural traditions, and how they continue to be undermined by Enlightenment thinking that supported the spread of capitalism as well as contributed to important gains in social justice—which are now being undermined by the techno-fascist characteristics of the digital revolution that is bringing all aspect of daily life under constant surveillance.

Overall, the above essays advance eco-justice thinking partly by introducing examples of non-eco-justice ideas. The essays also highlight the importance of introducing the lived cultural patterns that need to be the focus of the students' attention if they are going to recognize the current misconceptions that have their origins in the print-based abstractions about individual autonomy, and human-entered world, and infinite progress. Ethnographies will provide students the basis for identifying sustainable and unsustainable daily practices as well as lead to a more complex understanding of the nature of traditions. As the forces of climate change begin to impact the economic system, such as we are now witnessing in the decline of the coal industry, and in the destruction of people's livelihoods resulting from floods and extreme weather, it will be necessary to consider what traditions need to be intergenerationally renewed and carried forward as we enter the ecological endgame. Perpetuating the abstractions of the print-based elites who have controlled how modernization was to be understood and practiced will be seen as not only irrelevant but as what needs to be challenged. Eco-justice thinking needs to bring an ecological and culturally informed perspective that will lead to the recognition that the West's industrial revolution, and now the digital revolution, represent ecologically unsustainable detours from the local traditions of face to face and intergenerational decision making and mutual community support that have been the mainstay of human history.

❖ ❖ ❖

Chet Bowers is a semi-retired professor who continues to write on educational reforms that address the cultural roots of the ecological crisis. Among his twenty-four published books are two on technology: *Let The Eat Data* (2000) (translated into Japanese and Chinese) and *The False Promises of the Digital Revolution* (2014) and *Digital Detachment: How Computers Undermine Democracy* (2016).

References

Bateson, G. 1972. *Steps to an Ecology of Mind*. New York: Ballantine Books.

Berry, W. 2000. *Life is a Miracle: An Essay Against a Modern Superstition*. Berkeley, CA: Counterpoint Press.

Bowers, C. 2000. *Let Them Eat Data. How Computers Affect Education, Cultural Diversity, and the Prospects of Ecological Sustainability*. Athens, Georgia. University of Georgia Press.

_____. 2011. *Perspective on the Ideas of Gregory Bateson, Ecological Intelligence, and Educational Reforms*. Eugene, Or.: Eco-Justice Press.

_____. 2012. *University Reform in an Era of Global Warming*. Eugene, Or.: Eco-Justice Press.

_____. 2012. *The Way Forward. Educational Reforms that Focus on the Cultural Commons and the Linguistic Roots of theEcological/Cultural Crises*. Eugene, Or.: Eco-JusticePress.

_____. 2013. *In the Grip of the Past: Educational Reforms that Address What Should Be Changed and What Should Be Conserved*. Eugene, Or.: Eco-Justice Press.

_____. 2014. *The False Promises of the Digital Revolution: How Computers Transform Education, Work, and International Development in Ways that are Ecologically Unsustainable*. New York: Peter Lang.

_____. 2016. *Digital Detachment: How Computers Undermine Democracy*. New York: Routledge Press.

_____. 2016. *A Critical Examination of STEM: Issues and Challenges*. New York: Routledge Press.

_____. 2016. *A Historical Detour That May Be Fatal: What We Can Learn from the Luddite's Community-Centered Approach to Technology*. Eugene, Or.: Eco-Justice Press.